MW00436795

"Criminal justice can contribute t(
A significant share of greenhouse gas emissions is associated with conduct amounting to violations of existing criminal law. Targeting climate change by enforcing criminal law can be extremely efficient. It can be done on the basis of existing laws, through existing institutions and with minimal additional cost. Peter Carter and Elizabeth Woodworth's book is a timely and important contribution to the debate regarding how criminal prosecutions, both at the national and international level, could be used to repress and deter climate damaging conduct at a large scale and on a lasting basis." **—Reinhold Gallmetzer**
Appeals Counsel, International Criminal Court
Founder, Center for Climate Crime Analysis

"This book is more than timely — it constitutes a last call on the fast approaching calamity for humanity and for nature. The current CO2 rise rate of 2 to 3 parts per million [per year] is the fastest recorded over the last 66 million years. Given amplifying CO2 and methane feedbacks from warming oceans, drying land and [melting] permafrost, without effective efforts at CO2 draw-down, it is hard to see how the rise in global temperature can be halted, with tragic global consequences." **—Dr. Andrew Y. Glikson**
Earth and Paleoclimate Scientist
Australian National University

"Peter Carter and Elizabeth Woodworth build a damning case against fossil fuel companies and their political agents, showing that discounting of global warming in pursuit of short term profit is a crime against humanity. In this excellent, well-researched book, the authors map the global effort needed to survive the challenge of global warming; and perhaps to emerge wiser for our labors. *Unprecedented Crime: Climate Science Denial and Game Changers for Survival* is an indispensable read for the citizens and policy makers who will fight for civilization's endurance and advancement." **—Lawrence Torcello**
Associate Professor of Philosophy
Rochester Institute of Technology

"At this mind-blowing moment in time, when humanity faces irrevocable climate tipping points, we need all citizens to call out the climate emergency and take action. This book is a call to arms."

—Mary Christina Wood
Professor of Law and Faculty Director
Environmental and Natural Resources Law Center
University of Oregon

"Green criminologists are increasingly focusing attention on the critical issue of corporate and state crimes related to climate change. This important book makes a significant contribution to those efforts by pulling together a wealth of material to examine one such form of climate criminality, the unprecedented crime of climate science denial. This is a timely analysis given the current political crisis in the U.S., with excellent descriptions of some Game Changers for Survival."

—Ronald C. Kramer
Professor of Sociology
Western Michigan University

"This book makes an important contribution to the public debate about climate change because it effectively explains why the climate change disinformation campaign funded by some fossil fuel companies and free-market fundamentalist foundations should be responded to as an unprecedented virulent crime against humanity. Although the book makes clear that enormous harms to life and ecological systems are already attributable to this disinformation campaign, if the obstructionist power of the disinformation campaign can be neutralized, policy responses are now available that could effectively minimize future climate-induced catastrophic harms."

—Donald A. Brown
Scholar In Residence and Professor
Sustainability Ethics and Law
Widener University Commonwealth Law School
Contributing Author, Intergovernmental Panel on Climate
Change, 5th Assessment

UNPRECEDENTED CRIME

Climate Science Denial and Game Changers for Survival

by

DR. PETER D. CARTER
ELIZABETH WOODWORTH

Foreword by James E. Hansen

CLARITY PRESS, INC.

ISBN: 978-0-9986947-3-3
EBOOK ISBN: 978-0-9986947-4-0

In-house editor: Diana G. Collier
Cover design: R. Jordan P. Santos

*Some of the graphs herein have been modified for visual clarity. Explanatory titles for some graphs have been supplied at the top of the image by the authors.

Library of Congress Cataloging-in-Publication Data

Names: Carter, Peter D. (Environmentalist), author. | Woodworth, Elizabeth, 1943- author.
Title: Unprecedented crime : climate science denial and game changers for survival / by Dr. Peter D. Carter and Elizabeth Woodworth ; foreword by James E. Hansen.
Other titles: Climate science denial and game changers for survival
Description: Atlanta, GA : Clarity Press, Inc., 2018. | Includes bibliographical references and index.
Identifiers: LCCN 2017056416 (print) | LCCN 2017061123 (ebook) | ISBN 9780998694740 | ISBN 9780998694733 (alk. paper) | ISBN 9780998694740 (alk. paper)
Subjects: LCSH: Climatic changes--Social aspects. | Climatic changes--Political aspects. | Intergovernmental Panel on Climate Change.
Classification: LCC QC903 (ebook) | LCC QC903 .C3668 2018 (print) | DDC 363.738/74561--dc23
LC record available at https://lccn.loc.gov/2017056416

Clarity Press, Inc.
2625 Piedmont Rd. NE, Ste. 56
Atlanta, GA. 30324 , USA
http://www.claritypress.com

TABLE OF CONTENTS

PART II:
GAME CHANGERS FOR SURVIVAL

To all of today's children, everywhere

ACKNOWLEDGEMENTS

We begin by acknowledging the innumerable people behind the scenes in laboratories and government agencies such as NASA and NOAA who are working hard to document and report the growing phenomenon of climate change.

We next thank the climate scientists who are analyzing the data and reporting their findings in peer-reviewed journals that are often overlooked by the media. We thank those in particular whose work we have cited herein.

We would also like to thank the climate scientists, sociologists, criminologists, lawyers, and philosophers who have put time aside to review this book prior to publication. In particular, we thank Dr. James Hansen, who kindly found the time to write the Foreword on the plane home from COP23 in Bonn, Germany.

Peter's wife, Julie Johnston, of GreenHeart Education, did a magnificent job of proofreading the entire manuscript, and aided the writing of the book with many thoughtful suggestions.

Diana Collier of Clarity Press has been a steady, supportive, resourceful and resilient publisher.

Finally, we acknowledge the young people who have been trying for six years to bring the Our Children's Trust court case to trial. They deserve what they are seeking from the U.S. government: the right to a safe and stable climate for their lives and their own unborn children.

CRIMINALITY, INDEED

Late in 1981, or perhaps early 1982, Wally Broecker invited me to help organize a symposium on Earth's climate change. I worked with Taro Takahashi, Wally's close colleague. This was to be a Ewing Symposium, a continuation of a series of such meetings on fundamental topics in Earth Sciences, named for Maurice Ewing, the founder of Lamont Doherty Geophysical Observatory (LDGO). LDGO was and is a scattering of buildings, with offices, laboratories and field equipment, in a wooded area overlooking the Hudson River from the palisades, about 25 miles north of New York City, just over the New Jersey border into Palisades, New York.

The peaceful appearance of LDGO belied an underlying intellectual ferment stirred by the desire to unlock an understanding of the workings of planet Earth. Although Robert Jastrow, the founding director of the NASA Goddard Institute for Space Studies near the Columbia University campus, liked to boast that the concepts of plate tectonics, popularly continental drift, were hatched at a geophysics conference at GISS, most of the advances in understanding were achieved by the people at laboratories such as LDGO.

Taro and I were to assemble a diverse group of scientists with different specialties bearing on an understanding of the ocean, atmosphere, Earth's climate history, and human-made factors that were driving climate change. From a personal perspective, it was an illuminating activity, as we heard from a wide range of experts. GISS colleagues and I contributed a paper in which we showed that climate sensitivity to natural or human-made forcings could be extracted by comparing glacial and interglacial climate states, provided that we knew the change of global temperature between those periods, the change of atmospheric composition and the change in size of the polar ice sheets.

Perhaps the most remarkable presentation of the entire symposium was the dinner speech by E. E. David, Jr., titled "Inventing the Future: Energy and the CO2 'Greenhouse' Effect". David, the President of Exxon Research and Engineering Company, properly focused on the characteristic of the climate system that makes human-caused global warming so dangerous and such a challenge for society. Specifically, David said, "The critical problem is that the environmental impacts of the CO2 buildup may be so long delayed. A look at the theory of feedback systems shows that where there is such a long delay the system breaks down unless there is anticipation built into the loop.»

Delayed response of the climate system, caused by the great thermal inertia of the ocean and the slow response of ice sheets to warming, creates the possibility that we could hand young people a planet undergoing changes that would be out of control. The "anticipation" that David spoke of, i.e., the action needed to avoid "system breakdown", would be to develop carbon-free energy sources, such as improved renewable energy and nuclear power. The fossil fuel companies would need to become energy companies—clean energy companies.

Instead, Exxon and the fossil fuel industry moved in the opposite direction: they invested enormous funds into development of unconventional fossil fuels, such as tar sands and shale oil. It required decades and many billions of dollars, but they successfully developed hydro-fracking to recover vast quantities of hard-to-reach hydrocarbons.

Something went wrong. The fossil fuel industry understood the situation. The government understood the situation. Science had informed the government and industry about the risk of continuing with business-as-usual. The system could be pushed beyond a point-of-no-return.

And so we see it coming to pass.

Peter Carter and Elizabeth Woodworth make an overwhelming case that the public, especially young people, are victims of "Unprecedented Crime". And the fossil fuel industry, they explain, are not the only perpetrators. There has been extensive collusion and denial.

Fortunately, Carter and Woodworth do much more than expose the crimes against humanity—they also present actions that people can take to alleviate the consequences for today's public and for future generations.

The most fundamental requirement for moving to clean carbon-free energies is a rising carbon fee collected from fossil fuel companies, so as to make the price of fossil fuels pay their costs to society, including the costs of air pollution, water pollution and climate change. If these funds are distributed uniformly to the public, most people will come out ahead, so they will allow the carbon fee to continue to rise. Products based heavily on fossil fuels will become more expensive and lose out. Economists agree that this is the most efficient way to phase down fossil fuel use, and the only plausible way to get global emissions to decline rapidly. Carter and Woodworth make this essential case for a carbon tax in conjunction with the vital case for ending fossil fuel subsidies worldwide.

The authors have provided an excellent overview of the CO2 crisis that can help to educate the public on both the roots of the climate change problem and on many of the solutions that already exist to curb it.

These solutions urgently await our political will and determined implementation.

—Dr. James E. Hansen
November, 2017

INTRODUCTION

Since the United Nations Paris conference in late 2015, climate change indicators have escalated so quickly that an emergency response is imperative if civilization is to avoid breakdown and eventual collapse.

This book takes an unusual approach to the entrenched failure of governments and the media to act decisively and effectively to drastically curb CO_2 emissions.

The fact that no emergency response has been mounted by national governments is a crime against humanity and indeed all of life.

To continue with business-as-usual at this late date is to knowingly, and therefore deliberately, compound this crime.

In March 2017, Noam Chomsky described the plight of human civilization as "racing towards the precipice."[1]

Chomsky cited the Doomsday Clock, which was invented by the *Bulletin of the Atomic Scientists* in 1947 to reflect the threat of global catastrophe posed by nuclear warfare. (The clock was started at 7 minutes before midnight.)

In 1953, when the U.S. and the Soviet Union began testing hydrogen bombs, the clock was set at its lowest historical point: two minutes before midnight.

In 2007, the clock was redesigned to include the human-caused catastrophe posed by climate change.

In January 2017, the doomsday clock was set to its lowest point since 1953, at two and a half minutes before midnight. The cause: "the rise of 'strident nationalism' worldwide... and the disbelief in the scientific consensus over climate change by the Trump Administration."

This rush towards the precipice has become much more serious since 2015, as shown by shocking rises in global temperatures and atmospheric greenhouse gases.

For example:

- For several days in April-May 2017 atmospheric CO_2 reached an unprecedented level of over 410 parts per million. Atmospheric CO_2 in 2015 and 2016 had increased by an unprecedented 3 ppm in one year, described by the National Oceanic and Atmospheric Administration (NOAA) as "explosive," and 100 to 200 times faster than the natural CO_2 increase that has ended ice ages over the past million years.
- Record sea surface temperature increases caused an unprecedented mass bleaching of the Great Barrier Reef two years in a row (2016 and 2017), making it likely that half to two thirds of the Reef will never recover.
- The Arctic summer sea ice continues its free-fall decline with three unprecedented heat waves during the 2016-2017 winter. The Arctic is now emitting more CO_2 and methane than it absorbs – a long-dreaded Arctic feedback."[2]
- A September 2017 research paper in *Science* has shown that widespread deforestation and degradation have switched tropical forests from being a carbon sink to a net carbon source, adding new urgency to the critical need for aggressive efforts to reduce greenhouse gases.[3]

The scientific prediction of these events goes back to the 1980s, and we have simply watched them unfold as forecast. At this late date, the survival of civilization will depend on which of two responses to the crisis will win out:

1) The governments of high-emitting countries, and the corporate media, continue their leadership void by failing to educate the public or to take immediate decisive action to curb emissions, even as climate change indicators soar beyond expectation.

2) Implementation of the scores of innovative solutions to curb the crisis that have been devised by other states, regional and municipal governments and thousands of businesses, native communities, NGOs and citizens.

Given this availability of solutions (which are largely absent from the North American media), the political, media, and moral failures to act decisively have now become *willful crimes against life itself.*

This conscious refusal of the dominant players–government, corporations and the media–to *act with purpose* helps to answer the questions, "Where is the human outcry for earth's life-support? Why have we failed to seize control of our survival?"

Trappist monk Thomas Merton explored a similar mystery in the 1960s. During the nuclear madness of the Cold War, he coined the term "the unspeakable" to describe a vacuum that can be utterly void of compassion and responsibility.

This moral abyss is still very much alive within the deep military-industrial state that Eisenhower warned about in 1961. The public needs to acknowledge and understand this nihilistic mindset, which since the 1980s has played a major role in hiding from society the truth about the climate crisis.

Pope Francis, in his 2015 *Encyclical Letter* – which was backed by all the major world faiths – referred to climate change as "a sin against God."

Following the Pope's declaration, the UN Paris climate summit was signed by 195 countries – yet astonishingly our North American national governments persist in activities of deep denial, as they rush ahead with new pipelines.

And incredibly, against years of IMF and World Bank pressure to phase them out,[4] governments continue to subsidize fossil fuels to the globally suicidal extent of trillions of dollars worldwide. It is as if our political leaders had no concern for our children's future. Simply stopping these subsidies would be an instant game changer.

Another powerful strategy is legal action against governments for the crime of omission to protect the right to life of their populations – a public trust duty that dates back to Roman times and early British common law.

This book is written for both general readers and informed readers. It is presented in two parts: "Crimes Against Life and Humanity," and "Game Changers for Survival." It will present:

1) a global overview of unprecedented extreme weather events in 2017;

2) the ongoing political and media efforts to *suppress climate change as a crisis*: by denial, by under-reporting solutions to it, and by fixating on adaptation to daily extreme weather events while failing to urge radical emissions reduction;

3) the emerging and largely unreported opportunities for major sectors of the economy to transition to renewable energy, while increasing jobs and profits;

4) a special chapter called "Mission Impossible," which introduces entirely new thinking on how powerful and transformative action can begin immediately;

5) a special Science Appendix so that the book can be appreciated by both the general reader and the climate expert; this appendix shows why the rapidly increasing CO_2, methane, and nitrous oxide concentrations actually constitute a climate emergency. The numbers will be a shock for informed readers because they have scarcely been reported beyond the National Oceanic

and Atmospheric Administration (NOAA) data releases.

Despite all obstacles, we are encouraged to see powerful game changers on the horizon. For one, the markets are forging ahead to bring wind and solar energy into mainstream dominance over fossil fuels, and in 2016 renewable energy accounted for two-thirds of the new power added to the world's grids, with solar power the fastest growing.[5]

Another game changer is the *Subnational Global Climate Leadership* (alias *Under 2°C*), whose signatories represent 1.2 billion people on six continents, and 39% of the global economy. Its 187 jurisdictions pledge "emission reductions for a trajectory of 80 to 95 percent below 1990 levels by 2050."[6]

Completely blacked out by the media since its inception in 2015, this outstanding initiative has been called a game changer by former UN Secretary-General, Ban Ki-moon.

Predictably, many such survival responses are being downplayed and ignored by the North American corporate media, which, although it now reports *incessantly and superficially* on climate change, is still criminally tied to the profits and employment generated by oil, natural gas, and coal. It has seldom if ever reported on the emergency climate mobilization movement, which has climate scholars and writers on its advisory board.[7] We define this willful, methodical blocking of vital survival information as an unprecedented crime against life on the planet.

It is very late – but not too late – to recognize and address the crime of ecocide in the form of business-as-usual. There is still time for the planet to recuperate, but we must start emergency life-saving measures to reduce CO_2 emissions to near-zero. (These measures are explored in the chapters "Game Changers in Technology & Innovation" and "Mission Impossible.")

The truth embedded in climate science is unstoppable, and solutions whose time has come simply need to be elevated into general consciousness and translated into government action at all levels, which is a central purpose in writing this book.

Endnotes

1   "Racing To The Precipice: Prof. Noam Chomsky" (March 2017) (https://www. youtube.com/watch?v=TK0R_06zOOY).

2   NOAA, *Arctic Report Card: Update for2016* (http://www.arctic.noaa. govReport-Card/Report-Card-2016).

3   "New measurements show widespread forest loss has reversed the role of tropics as a carbon sink," *phys.org*, 29 September 2017.

4   IEA, OECD and World Bank, Joint Report: *The Scope of Fossil-Fuel Subsidies in 2009 and a Roadmap for Phasing Out Fossil-Fuel Subsidies*, 2010 (https:// www.oecd.org/env/cc/46575783.pdf).

5   Adam Vaughan, "Time to shine: Solar power is fastest-growing source of new energy," *The Guardian*, 4 October 2017.

6   Subnational Global Climate Leadership Memorandum of Understanding (http://under2mou.org/).

7   The Climate Mobilization (http://www.theclimatemobilization.org/advisoryboard).

PART I

CRIMES AGAINST LIFE AND HUMANITY

Chapter One

EXTREME WEATHER AROUND THE WORLD

Early Climate Warnings and Their Suppression

Former NASA scientist Dr. James Hansen has been called "the father of climate change awareness." In the 1980s, he was the first scientist to blow the whistle on atmospheric CO_2 pollution, and has fought hard ever since to prevent the catastrophe he knew was coming.

Hansen, who had been studying changes to the climate since the 1970s, warned a Congressional committee in 1988 about the "greenhouse effect." There was a 99% certainty, he said, that heat-trapping gases released into the atmosphere were causing global warming, and that the likelihood of extreme weather events was growing steadily.

Following his testimony, *The New York Times* reported that "humans, by burning of fossil fuels and other activities, have altered the global climate in a manner that will affect life on earth for centuries to come."[1]

Things started getting difficult for Hansen. The next year the White House altered his testimony, and NASA appointed an overseer to vet what he said to the media. Any suggestion that fossil fuels should be reduced was discouraged as being political and unscientific.

In June 2006, Hansen appeared on *60 Minutes,* saying that the Bush White House had edited federal press releases to tone down the danger of climate change, and that he was not able to speak freely without backlash from government officials.

Twenty years after his Congressional testimony, Hansen made a public statement in 2008 that the world was now in a state of planetary emergency. He again warned emphatically that more warming was in the pipeline and would trigger irreversible tipping points that would affect life on earth for decades and centuries to come.[2]

Hansen, now in his 70s, has been arrested three times outside the White House for protesting mining and the Keystone XL pipeline.

In spite of his warnings, for the past 25 years a constant, increasingly aggressive campaign has been waged to deceive the public into thinking that atmospheric greenhouse gas pollution is nothing to worry about. For years this campaign, initiated and driven by US-based multinational fossil fuel corporations, has managed to control the prevailing view of the United States Congress on climate.

As a result, we are now in a planetary emergency playing out in the form of increasingly regular and devastating extreme weather events, with no effective plans to halt the greenhouse gas emissions that are deadly to this and future generations.

Understanding Why the Emergency is "in the Pipeline"

It is vital to understand what James Hansen meant when he said that more warming was in the pipeline, because herein lies the invisible dynamic that will increasingly propel the emergency.

In 2008 and again in 2012 James Hansen made a public statement that we had reached a point of planetary emergency because of today's extreme impacts and the crucial fact there is more unavoidable warming to come, which is in the pipe.

Most of the added greenhouse gas heat gets stored in the oceans. This results in a delayed full surface warming response, known as climate inertia or the ocean heat lag, caused by the thermal inertia of the oceans. We found this has been recognized by the Intergovernmental Panel Climate Change from 1990.

Thus we say that more future global surface warming is "committed" or "locked in" or "in the pipeline" as James Hansen has been warning for years.

In the assessment of all global climate change impacts, therefore, the impacts and risks should be determined by the committed or locked in degree of climate change today, not only by the degree of global warming that we currently see around us.

It is not possible to understand today's emergency for today's children and all future generations without understanding that the great store of ocean heat will be stoking the climate system as we move forward. This is why 195 nations in Paris agreed in 2015 to try to limit global warming to 1.5°C above pre-industrial times.

We turn now to the extreme weather events that illustrate the dire state of the climate, the land, and the oceans.

The Tomahawk wildfire burns through the night on the northeast portion of Marine Corps Base Camp Pendleton, Calif., May 16, 2014. The fire has burned more than 6,000 acres, forcing evacuations of housing areas on base and various schools on and off base.
U.S. Marine Corps photo by Cpl. Orrin Farmer

Extreme Weather, 2017

In January 2017, world climate monitoring centers reported that global temperatures in 2016 had broken records for an unprecedented third year in a row.

In September 2017, the US National Oceanic and Atmospheric Administration (NOAA) reported that 2017 had the second highest first half of a year (January–August) temperature increase, barely behind the record year of 2016.

NOAA said this near-record warmth is especially remarkable, given the lack of an El Niño event in 2017. This makes 2017 almost certain to be the planet's warmest year on record without an El Niño, and hence the four warmest years to be 2014 through 2017.

The full global reality of how 2017 played out is stunning.

Hurricanes

To most people living in the Global North,[3] climate change may have seemed – until 2017 – like a shadow on a distant horizon.

Although the US has suffered extreme floods, heat waves and drought over the past few years, and although reports have persisted of small islands sinking in the southern oceans, of heat waves in Europe, and of monsoons and failed crops in Asia, alarm has remained far from most Western minds.

But during the summer of 2017, the whole world looked on as survivors in Texas, Florida, Puerto Rico, Dominica, and the U.S. Virgin Islands were forced to rebuild their lives following three brutally punishing hurricanes.

Hurricane Harvey (Category 4, August 17 – September 3) unleashed the most destructive rainfall in US history – more than 49 inches in some areas of Houston, with floods displacing more than 30,000 people. Meteorologists referred to "unprecedented" 130 mile-per-hour winds that caused 82 deaths in the U.S. and one death in Guyana. Texas and Louisiana, soaked with 27 trillion gallons of water, will take years to recover.

Hurricane Irma, a Category 5, was the most intense Atlantic hurricane to strike the United States since Katrina struck in 2005, killing 44 people in the Caribbean and 58 in the U.S. Irma's 185 mph maximum winds lasting for 37 hours was the longest intensity maintained by any cyclone in the world – the previous record of 24 hours was held by the Pacific's Super Typhoon Haiyan in 2013.

Fueled by warm surface ocean waters, Irma caused widespread catastrophic damage throughout its long life (August 30 - September 15), particularly in the northeastern Caribbean, Cuba and the Florida Keys. According to the National Hurricane Center, Irma was the strongest hurricane ever recorded to form in the Atlantic Ocean outside of the Gulf of Mexico and Caribbean Sea.

On the heels of Irma came Maria, also a Category 5. Forming September 16, Maria made devastating landfall in Puerto Rico September 20, causing deaths across Guadeloupe, Haiti, Puerto Rico and the Dominican Republic. Puerto Rico's electrical grid was totally wiped out, leaving the entire island without power. Then, after 16 inches of torrential rains, the 90-year-old Guajataca Dam was damaged and its emergency spillway breached, causing tens of thousands to be evacuated.

The New York Times reported on September 19 that "it will almost certainly be the most expensive hurricane season on record in the United States." Compared to Katrina's $143.5 billion, Harvey and Irma together cost $290 billion.

The fury of 2017's unprecedented series of Atlantic hurricanes was energized by the ever-warming ocean surface, because of the intensifying greenhouse gas blanket above it (see the Science Appendix for further information).

Heat Waves and Wildfires

Background: Before 2017, the world had already suffered widespread heat catastrophes. The European heat wave of 2003 led to the hottest summer on record since 1540 and killed more than 70,000 people.

In 2010, the blistering Russian heat wave was even more extreme, covering 400,000 square miles, about double that of 2003. Flames ravaged 1.25 million hectares of land, and 22 million hectares of crops were destroyed. Munich Reinsurance Company estimated that 56,000 people had perished from the combined effects of smog and heat. Moscow's health department reported that deaths had nearly doubled in the city to about 700 a day.

The severe and widespread 2012 North American heat wave drove a record-shattering drought that caused massive crop failures throughout the Midwest, affecting 80% of the country. July 2012 was the hottest month in U.S. history up until that time.

In 2017, extreme heat and record temperatures were recorded across the western United States and British Columbia in June and July. Portland, Oregon recorded a new daily temperature high of 99°F. In San Francisco, temperatures broke all-time records over Labor Day, reaching a sweltering 106°F, and the San Diego region of the Pasqual Valley was even higher at 108°F. In July, Death Valley in California set a new record for hottest month in US. history, with an average temperature ofof 107.6°F, which is also a record for the western hemisphere.

In a fire season 78 days longer than in 1970, wildfires ignited the Pacific Northwest, the Sierras, Los Angeles, and much of the western United States and Canada. Breaking 100-year heat records, British Columbia suffered the largest burn area in history (nearly 3 million acres), with the highest number of evacuees (est. 45,000), and the largest single fire ever recorded in the province. A state of emergency, declared July 7, was extended three times until September 1.

Chile declared a state of emergency as its worst wildfires in history ravaged the country in January. During an intense heat wave of 35°C with strong winds, more than 4,000 firefighters and equipment were brought in from 23 countries to fight the 130 blazes.

In June, Portugal suffered a catastrophic series of wildfires that trapped and killed 64 people. Preceded by an intense heat wave, with much of Portugal seeing temperatures in excess of 40°C (104°F), 156 wild fires erupted across the country.

In the United Kingdom, heat wave alerts were issued throughout the summer. June temperatures were the hottest in 40 years with Heathrow Airport registering 34.5°C (94.1°F).

In the first week of August, authorities in eleven European countries warned people to take precautions amidst the region's most intense heat since the deadly heat wave of 2003. Temperatures exceeded 40°C (105°F) – 10-15°C higher than is normal for the time of year. This followed a spate of extreme temperatures in July that fuelled major wildfires, severe drought and damaged crops in Italy and Spain.

Arab News reported in early July that in central and eastern parts of Saudi Arabia the temperature had reached 53°C for the first time ever. The government banned working under the sun from noon until 3 p.m. from June until September.

Scientists were amazed that even in Greenland a large wild grass-fire burned for two weeks in late July only 40 miles from the ice sheet. Dangerously compounding the situation, the soot turns the ice sheet black, causing faster melt.

The heat wave reached all the way to Siberia, causing temperatures up to 37°C, with Moscow enduring July heat 2.7°C above normal.

Most of China was hit by extreme heat, with Shanghai recording the hottest day ever of 40.9°C (105 °F). Heat waves struck India earlier than normal in May and June. Record temperatures melted road asphalt. NOAA reported that July in Africa and the Pacific Ocean islands was the hottest since continental records began in 1910.

On February 12, 2017, record heat was recorded in many regions of Australia, including a temperature of 46.6°C (115.9°F) in the coastal city of Port Macquarie, New South Wales. Two days earlier, the average maximum temperature across all of New South Wales had hit a record-setting 42.4°C (108.3°F) – a record broken the next day when it rose to 44.0°C (111.2°F).

Rainfall and Flooding

Although monsoons hit South Asia every year between June and

September, according to the BBC the 2017 monsoon season was far worse than average. Major monsoons and floods left a third of Bangladesh and huge swaths of India, Pakistan and Nepal underwater. The International Federation of Red Cross and Red Crescent Societies called the South Asian floods one of the worst regional humanitarian crises in decades. By early September 1,288 people were confirmed dead. More than 45 million were impacted, and ruined farmland reduced food supplies.

In China, floods began in early June, impacting 15 million people in 10 provinces and destroying 18,000 homes. The State Flood Control and Drought Relief Headquarters reported July 2nd that water levels in more than 60 rivers in southern China were above the warning levels due to sustained rainfalls for days on end.

Sudden floods in Iran and Livorno, Italy caused the deaths of 40 people and 8 people respectively.

In South America, unusually heavy rains in late February pummeled the Santiago region of Chile, where flooding and landslides killed at least eighteen people. In Brazil, early March rains measuring up to 4.3 inches in 24 hours caused flooding across the state of Rio Grande do Sul, killing 2 people and injuring 70.

Northern California saw its wettest winter in almost a century, with landslide damage to roads and highways estimated at over $1 billion.

In Canada, floods hit southern Quebec on May 3 following excessive rainfall. The government reported 2,426 flooded residences, 2,720 people forced from their homes, and 146 impacted municipalities. Montreal and Laval declared states of emergency.

Since 1995 the IPCC Assessments have confirmed these changes with increasing certainty, and have recorded the actual increases in extreme weather events.

Conclusion

In 2017 climate change landed on the doorsteps of millions of people in the developed world. The images of what climate change looks like no longer belong to the future.

And finally the weather reports are pointing non-stop to climate change as the cause of sensational, cascading worldwide emergencies. They talk about blackened fire crews, drowning houses, collapsing bridges and flooded hospitals. They talk about raising billions upon billions to rebuild. Indeed they talk about everything but the root problem – and the emergency measures needed to stop it from getting worse.

That would mean raising a subject being denied by our governments: the need to immediately and rapidly slash CO_2 emissions – and, according to the IPCC, methane and nitrous oxide to "near zero" as well.

This was stated very clearly in a 2015 World Meteorological Organization press release, in which Secretary-General Michel Jarraud said:

> Every year we say that time is running out. We have to act NOW to slash greenhouse gas emissions if we are to have a chance to keep the increase in temperatures to manageable levels.[4]

This will entail a fundamental transformation in business-as-usual, a subject seldom mentioned in the media, as we shall see in Chapter Four. For now we will simply quote British journalist Ellie Mae O'Hagan, who wrote in September 2017:

> A new campaign could centre on the demand for governments to meet the 1.5°C target, emphasising how dire the consequences will be if we don't.
>
> Could the language of emergency work? It has never been tried with as much meteorological evidence as we have now, and we've never had a target as clear and unanimous as the one agreed in Paris. The one thing I know is that the events of the last few months have changed the game, and this is the moment to start debating a new way to talk about climate change. It may be that if the time for a mass movement is not now, there won't be one.[5]

Endnotes

1   Philip Shabecoff, "Global Warming Has Begun, Expert Tells Senate," *The New York Times*, 24 June 1988.

2   Jim Hansen, "Global Warming 20 Years Later: Tipping Points Near," National Press Club and House Select Committee on Energy Independence & Global Warming, 23 June 2008 (http://www.columbia.edu/~jeh1/2008TippingPointsNear_20080623.pdf).

3   The "Global North" refers to wealthy industrialized countries, including the United States, Canada, Western Europe, Russia, Israel, Australia, New Zealand, and developed parts of Asia (China, Taiwan, Singapore, Hong Kong, South Korea, and Japan).

4   World Meteorological Organization, "Greenhouse Gas Concentrations Hit Yet Another Record," 9 November 2015.

5   Ellie Mae O'Hagan, "Climate optimism has been a disaster. We need a new language – desperately," *The Guardian*, 21 September 2017.

Chapter Two

SCIENCE BETRAYED: THE CRIME OF DENIAL

In this chapter we will argue that the denial of climate change science by corporations, and by the institutions that govern and inform society, is the worst ever crime against humanity.

The Global Foundation of Climate Science

The Intergovernmental Panel on Climate Change (IPCC) produced its first assessment in 1990, a pivotal year for the understanding of global warming. This assessment informed the UN negotiations that resulted in the signing, by all countries, of the *UN Framework Convention on Climate Change* in 1992.

With the signing of this document, the world as a whole learned of, and recognized, the extreme global dangers of continued, unmitigated greenhouse gas (GHG) emissions.

The IPCC climate change assessments include input from policymakers representing all the world's governments – in addition to leading scientists from all world regions. Indeed the IPCC assessment process is unique in that policymakers who sit on the Panel *do* represent all national governments.

These participants scrutinize the science reports line by line and *must agree on a final text in order for the assessments to be published.*

The original benefit of having world government policymakers on the Panel was that all governments would be made aware of the IPCC's important findings, which are published in the *Summaries for Policy Makers (SPMs).*

The assessments are conservatively understated, and their endorsements by all national governments makes the findings incontrovertible. Indeed many people have relied almost exclusively on the *SPMs* to understand the science, with the result that they are widely quoted. From the first 1990 IPCC assessment there has never been any doubt that human-sourced greenhouse gas emissions cause global warming.

The First IPCC Assessment (1990) said:

We are certain of the following:

- There is a natural greenhouse effect which already keeps the Earth warmer than it would otherwise be.
- **Emissions resulting from human activities** are substantially increasing the atmospheric concentrations of the greenhouse gases carbon dioxide, methane, and nitrous oxide.
- **These increases will enhance the greenhouse effect, resulting on average in an additional warming of the Earth's surface temperature.**
- The main greenhouse gas, water vapour, will increase in response to global warming and further enhance it. (Our emphasis)
- Carbon dioxide has been responsible for over half the enhanced greenhouse effect in the past, and is likely to remain so in the future.

"We calculate" are further to the left than the bullets up above:

- Atmospheric concentrations of the long-lived gases (including carbon dioxide and nitrous oxide) adjust only slowly to changes in emissions. Continued emissions of these gases at present rates would commit us to increased concentrations for centuries ahead.

- The longer emissions continue to increase at present day rates, the greater reductions would have to be for concentrations to stabilise at a given level.
- The long-lived gases would require immediate reductions in emissions from human activities of over 60% to stabilize their concentrations at today's levels; methane would require a 15-20% reduction."[1]

How quickly will global climate change?

The 1990 assessment calculated that *business-as-usual* emissions will result in a likely increase in global mean temperature of "about 4°C above pre-industrial before the end of the next [21st] century."[2] (Under this worst-case business-as-usual scenario the 2014 IPCC assessment projected a slightly higher global warming of 4.3°C by 2100.) It is important to bear in mind that while all the assessments make projections up to 2100, global warming does not stop at 2100.

The 1990 assessment also estimated that CO_2 emissions remain in the atmosphere 50-100 years.[3] In 2016, the U.S. Carbon Dioxide Information Analysis Center put the atmospheric lifetime of CO_2 higher, at 100 to 300 years.[4]

The science of "climate change commitment" was well-established and documented by the time of the 1990 assessment. As we saw in Chapter One, this science says that the degree of global climate disruption the world is experiencing today is already committed or "locked in" to more disruption in the future because of inertia in earth's systems.

The reason is quite simple: Increasing atmospheric carbon dioxide traps more and more heat in the earth's atmosphere, which causes a global energy imbalance because there is more energy coming in than escaping. Most of the heat goes straight to the oceans.

As a result, the planet will keep warming until it reaches a new balanced energy state – or equilibrium – with equal incoming and outgoing energy (when emissions have ceased, and the greenhouse gas blanket has stabilized). Thus the accumulating emissions will continue to drive temperatures up until after greenhouse gas emissions cease altogether, and equilibrium is later reached.

The National Research Council, in a long 2011 report (Climate Stabilization Targets), put it this way:

> Stabilizing atmospheric concentrations (of CO2) does not mean that temperatures will stabilize immediately. Because of time lags inherent in the Earth's climate, warming that occurs in response to a given increase in the concentration of carbon dioxide ("transient climate change") reflects only about half the eventual total warming ("equilibrium climate change") that would occur."[5]

In addition as the surface warming increases, it will be boosted by feedback emissions. An obvious example is increased forest fires that emit CO2, black carbon and some methane. Today's warming is 1.1°C but we must plan for 2.2°C being locked in.

Atmospheric CO2 Concentrations
NOAA Mauna Loa

2013 - Oct 2017

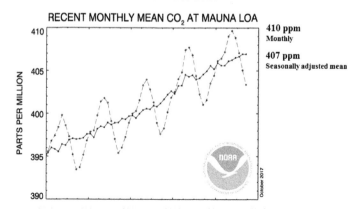

The IPCC estimates that if the carbon dioxide concentration in the atmosphere were to reach double its preindustrial level (which was about 280 ppm), the planet would warm 3°C, and 4.3°C up to a possible 5.4°C under business-as-usual by 2100.

In May 2017, weekly CO_2 concentrations at Mauna Loa registered a record 410.8 ppm before starting the seasonal decline each year as the green vegetation absorbs CO_2. The seasonally adjusted mean atmospheric CO_2 for September 2017 has increased to 407 ppm. This is the highest CO_2 in millions of years, and the rate of increase has also been unprecedented for at least a million years.

Atmospheric CO_2 concentration is still accelerating and faster than ever. This continues even though there is no more 2015-2016 El Nino ocean-warming effect (that tends to boost atmospheric CO_2), and even though fossil fuel CO_2 emissions stopped increasing in 2014.

This trend in atmospheric CO_2 concentration increase is on course with the worst case IPCC 2014 scenario (called RCP8.5), which leads to a best estimate warming from atmospheric GHGs of 4.3°C by 2100. However, the IPCC says it could be as high as 7.8°C by 2100 when including uncertainties such as amplifying feedbacks. Large feedback emissions are certain at 3°C.

Global Greenhouse Gas Emissions

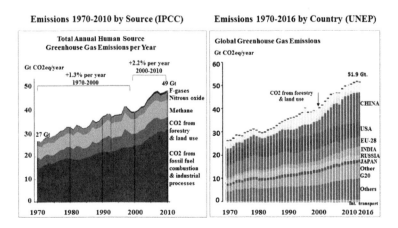

Emissions 1970-2010 by Source (IPCC)   Emissions 1970-2016 by Country (UNEP)

Source: IPCC 2014 WG3   Source UNEP GAP Report 2017

This means that government policies of high-emitting countries are risking a temperature of 7.8°C by 2100, far higher after 2100. These are not tolerable degrees of climate change for our species, nor most others.

Since 1990 global GHG emissions have been allowed to keep increasing fast and are now 30% higher than in 1990. From 1970 to 2000, emissions increased 1.3% a year, but 2.2% a year from 2000 to 2010.

The science has been consistent from the start that in order to stabilize GHG pollution impacts on the climate and oceans, CO_2 (now with methane and nitrous oxide as CO_2-equivalent from the 2014 IPCC) must be brought down to near zero:

- The First Assessment in 1990 reported that stabilizing atmospheric CO_2 at 350 ppm would take an immediate 60% cut in emissions, and that to reduce atmospheric CO_2 (very slowly) required zero carbon emissions.[6]
- The 2007 Fourth Assessment reported that "the most stringent scenarios could limit global mean temperature increases to 2-2.4°C above pre-industrial level. This would require emissions to peak by 2015 [at] the latest."[7]
- In 2014 the "near zero" required goal of all the long-lived greenhouse gase-was emphasized in the IPCC Fifth Assessment. Many experts now believe that near-zero should be reached by 2050 to avoid more than 1.5°C.
- Near-zero by 2050 is supported by 2016 research on the costs to society of interacting tipping points. To address these, the "optimal policy involves an immediate, massive effort to control CO_2 emissions, which are stopped by mid-century, leading to climate stabilization at less than 1.5°C above pre-industrial levels."[8]
- But our global commitments are not enough. In

March 2017, *Climate Interactive* estimated that the latest combined national emissions targets filed with the UN are unrealistic, and will lead to the disastrous temperature increase of 3.3°C by 2100, with a catastrophic risk up to 4.4C,[9] which would double after 2100 with climate system inertia and amplifying feedbacks.

- In a June 2017 letter to *Nature*, 60 eminent scientists, business leaders, economists, and others warned that we have three years to save the planet from the risk of grave climate risk.[10]

High-emitting governments are still declining to act upon the science: that saving the planet over the long term can only succeed with 100% fossil fuel conversion to clean energy, zero deforestation, and zero cement-making (burning limestone emits CO2). And these governments continue to ignore the will of global civil society.

In June 2014, the Climate Action Network International, representing over 90 organizations in over 100 countries, presented at the UN climate negotiations. They called for an end to the fossil fuel era: by cutting emissions to zero before 2050, and by an accelerated transition to a 100% renewable energy future aimed at a global warming limit of 1.5°C.

Accepting this near zero emissions reality actually makes mitigation simpler: All large GHG-polluting sources must and can be converted to non-GHG polluting processes – and for years we have known how.

The Difference between a 1.5°C World and a 3°C World (by 2100)

According to Climate Interactive in 2017, current emissions policies place the world on track for a 3.3°C (up to a possible 4.4°C) global temperature increase over pre-industrial times by 2100.

A 3°C world existed about three million years ago. At that time, there was virtually no Arctic sea ice, sea level was 20 meters higher, and forests extended across what is now tundra to the Arctic Ocean. According to the 2014 IPCC assessment, at 3°C, heat waves and droughts would cause a big drop in world food production affecting all major producing regions.

We are above 1.1°C already. The Paris Agreement and climate scientists[11] call for limiting global temperature increase to 1.5°C by 2100. NASA has projected the impacts of a 2-degree rise over a 1.5-degree rise:

- Heat-wave duration, rainstorm intensity and sea-level rise would increase by roughly a third (and sea level will continue rising long after air temperature is stabilized)
- There would be a disproportionately greater impact on basic crops
- Tropical coral reefs would be wiped out
- Fresh water in the Mediterranean area would be reduced by half.[12]

It is clear that we must avoid a catastrophic 2°C world and commit to a long term limit below 1.5°C.

It is also clear that for nearly 30 years the IPCC projections have been borne out by subsequent real-world observations. The IPCC consensus process,[13] the participation and knowledge of national governments, and the reporting of results to the world have all been well-conceived, conservative, and consistent. The decades of warnings from the IPCC are unassailable because they have been scrutinized and approved by policy-makers representing all world governments.

How can it be, then, that since the landmark international signing of the *UN Framework Convention on Climate Change* in 1992, global emissions have increased by 30%?

The Crime of Climate Change Denial

In asserting that climate science denial is the worst crime against humanity, we will first clarify our definitions.

Climate change denial has been led by industry disinformation, which according to Merriam-Webster is "false information deliberately and often covertly spread in order to influence public opinion or obscure the truth."

A crime against humanity is, according to the Oxford Dictionary, "a deliberate act, typically as part of a systematic campaign that causes human suffering or death on a large scale."

A Brief Look at the Origins of Denialism

In 2010 a landmark book, *Merchants of Doubt*, showed how a small group of prominent scientists with connections to politics and industry led disinformation campaigns denying established scientific knowledge about smoking, acid rain, DDT, the ozone layer, and global warming.[14]

Written by Dr. Naomi Oreskes, Harvard science historian, and NASA historian Erik Conway, *Merchants* was reviewed by Bill Buchanan of *The Christian Science Monitor* as "the most important book of 2010," and by *The Guardian's* Robin McKie as "the best science book of the year." It was followed by the 2014 documentary of the same name, also widely seen and reviewed.

The research showed how the disinformation tactics of the tobacco companies in the 1960s to undermine the scientific link between smoking and lung cancer served as a model for subsequent oil company tactics suppressing climate change science.

Following the U.S. Surgeon General's landmark report on smoking and lung cancer in 1964, the government legislated warning labels on cigarette packages. But a tobacco company executive from Brown & Williamson had a brainwave: people still wanted to smoke and doubt about the science would give them a ready excuse.

His infamous 1969 memo read: "Doubt is our product since it is the best means of competing with the 'body of fact' that exists in the minds of the general public. It is also the means of establishing a controversy."[15]

Tobacco industry executives never directly denied the mounting evidence that cigarettes were linked to lung cancer. Instead they stated publicly that the science was controversial. In this way they managed to delay regulation and lawsuits until the 1990s.

When the global warming science began to emerge in the 1980s, the oil industry employed the same deceptions. The whole focus was now on creating doubt in the minds of the politicians, the media and the public about whether we really know for sure that climate change is a problem. Doubt, as the tobacco industry had learned so profitably, delays action.

When the IPCC was formed in 1988 and began documenting and publicizing the impacts of climate change, the climate disinformation campaign grew more intense. Big Oil employed the same tactics, arguments, vocabulary, and PR firms that the tobacco companies had used to cast doubt on the dangers of smoking 25 years earlier.

The American Petroleum Institute convened a Global Climate Science Communications Team in 1998 to devise a plan targeting the media, schools, government officials, Congress, and other influential groups.

The team's mission, exposed in a leaked 1998 memo, was to initiate "a national media relations programme to inform the media about uncertainties in climate science; to generate national, regional and local media on the scientific uncertainties and thereby educate and inform the public, stimulating them to raise questions with policymakers." They said victory would be achieved when:

- Average citizens understand (recognise) uncertainties in climate science; recognition of uncertainties becomes part of the "conventional wisdom"
- Media "understands" (recognises) uncertainties in climate science
- Media coverage reflects balance on climate science and recognition of the validity of viewpoints that challenge the current "conventional wisdom"

- Industry senior leadership understands uncertainties in climate science, making them stronger ambassadors to those who shape climate policy
- Those promoting the Kyoto treaty on the basis of extent science appears [sic] to be out of touch with reality.[16]

A 2009-2014 study shows that climate change deniers promoting these uncertainties were prominently featured on CNN, MSNBC, Fox News, Fox Business, ABC, CBS, and PBS in a striking number of TV appearances – indeed three years after the publication of *Merchants of Doubt*.

These deniers included the non-climate scientists:

- Marc Morano (Bachelor PoliSci) from Climate Depot, 30 TV appearances
- Tim Phillips (Bachelor PoliSci) from Americans for Prosperity, 7 appearances
- Fred Singer (physicist) from the Science and Environmental Policy Project, 8 appearances
- James Taylor (lawyer), from the Heartland Institute, 8 appearances.[17]

Although these men lack credentials in climate science and have been widely exposed as imposters, the major cable TV and networks still give them credibility on their free media platforms.

The corporate media has thus given a relatively small group of science deniers with financial connections to the fossil fuel industry immense influence in sowing doubt on the scientific consensus of human-made climate change.

Climate Denial Propaganda and Influence Continue to Rise

In 2016 the Union of Concerned Scientists reported that "an in-depth analysis of eight leading fossil fuel companies finds that none of them has made a clean break from disinformation on climate science and policy." The companies included were ArchCoal, BP, Chevron, ConocoPhillips, Consol Energy, ExxonMobil, Peabody, and Shell.[18]

The industry has responded to the spotlight by intensifying propaganda through the agents below.

The Heartland Institute: In March 2017 the Heartland Institute began targeting the nation's 200,000 science teachers by mailing each a copy of its new book and DVD, "Why Scientists Disagree About Global Warming." The slick package stated that even if climate change were real, "it would probably not be harmful, because many areas of the world would benefit from or adjust to climate change."[19]

The non-profit National Center for Science Education, which monitors climate change education in schools, said the Heartland package contained little data but "was dressed up to look like science"; also that it was intended to confuse the 31% of teachers who tell their classes that the cause of climate change is still being debated.[20]

By 2017, the influence of the Heartland Institute (which had helped Philip Morris deny the risks of second-hand smoke in the 1990s) was at an all-time high, having advised the Trump administration on the appointment of climate change denier Scott Pruitt to head the U.S. Environmental Protection Agency (EPA). Heartland CEO Joseph Bast, a University of Chicago economics dropout, describes himself as a "global warming realist" and boasts that the White House continues to consult Heartland for "expert opinion," such as the need to dismantle the Clean Power Plan.[21]

Heartland enjoys a 7 million dollar contribution budget, much of which comes from the Koch Brothers, who are also key funders of the right-wing advocacy infrastructure, including the American Legislative Exchange Council (ALEC, a tax-exempt organization of conservative state legislators and private sector representatives), the Institute for Energy Research, Freedom Partners, and the State Policy Network (SPN).

The Koch Brothers: The multibillionaire industrialists Charles and David Koch are two of the most powerful people in the global oil industry, owning Koch Industries, a $100 billion conglomerate employing 100,000 people in 60 countries. They control 1-2 million acres of Alberta's tar sands.

In her 2016 best-seller, *Dark Money*, *New Yorker* journalist Jane Mayer chronicled the takeover of Republican politics by a massive infusion of Koch oil dollars over three decades. *The New York Times* reviewed it:

> The midterm elections of 2010 ushered in the political system that the Kochs had spent so many years plotting to bring about. After the voting that year, Republicans dominated state legislatures; they controlled a clear majority of the governorships; they had taken one chamber of Congress and were on their way to winning the other. Perhaps most important, a good many of the Republicans who had won these offices were not middle-of-the-road pragmatists. They were antigovernment libertarians of the Kochs' own political stripe. The brothers had spent or raised hundreds of millions of dollars to create majorities in their image. They had succeeded. And not merely at the polls: They had helped to finance and organize an interlocking network of think tanks, academic programs and news media outlets that far exceeded anything the liberal opposition could put together.[22]

The Real News ran a documentary later in the year, which interviewed Mike Casey of NextGen Climate Action: "The Kochs are a vertically integrated fossil fuel conglomerate, and they have a vertical integrated influence-peddling apparatus to go with it." Aliya Haq of the Natural Resources Defense Council concurred: "The Kochs' primary goal is to stop climate action at every turn."[23]

The Kochs are bigger than either of the Democratic or Republican parties, manipulate both, and are determined to keep the Senate Republican. In 2014 the Koch network spent $290 million to help Republicans gain control of the Senate.[24] By 2016 they had donated to 52% of the US Senate members and 33% of the Congress, and planned to spend $890 million on the 2016 election cycle.

A major focus of Koch money has been to ensure that no legislation is passed to curb the burning of fossil fuels. The brothers have gained pledges from 170 members of Congress that they will never support a tax on carbon, including some, such as Newt Gingrich, who reversed their earlier position of concern about climate change.

Republicans who refused to sign paid a heavy price. The Koch machine made an example of Republican Bob Inglis, who pledged allegiance but changed his mind and began to support climate science. Americans for Prosperity, central to the Koch network,[25] financed a candidate running against Inglis in 2010, causing him to lose his seat in South Carolina.[26]

When President Donald Trump backed out of the Paris Accord in June 2017, Jeffrey Sachs said in an interview, "the Koch Brothers have bought and purchased the top of the Republican Party…Trump is a tool in this. This is really Republican leadership…this is *not* Trump's fantasy world."[27]

Responsibly vocal scientists have paid a heavy price. Climatologist Michael Mann, Distinguished Professor of Atmospheric Science at Penn State University who originated the "hockey stick" analogy, received death threats and was brutally targeted in 2010 by the Koch's political action committee, which paid for Joe Barton, then Chair of the House Energy and Commerce Committee, to subpoena Mann's records and private emails, and to smear him in op-eds.[28]

Barton, a Texas lawmaker, had received at least $1.7 million in political contributions from oil and gas interests over the past two decades.[29] Congress was not conducting an evidence-based query, but a politically-driven attack on science designed to block climate action.

While attacking legitimate climate scientists, the Kochs were funding prominent pseudo-climate-scientists in other fields, such as aerospace engineer Willie Soon, who had been given a "courage award" by the Heartland Institute for calling global warming a hoax. Writes Suzanne Goldenberg in 2015:

> Over the last 14 years Willie Soon, a researcher at the Harvard-Smithsonian Centre for Astrophysics, received a total of $1.25m from Exxon Mobil, Southern Company, the American Petroleum Institute (API) and a foundation run by the ultra-conservative Koch brothers."[30]

ExxonMobil: In 2015, we learned from its own research that Exxon has known since 1980 that global warming is real.[31]

Kert Davies, former research Director of Greenpeace USA, revealed through ExxonSecrets.org that, meanwhile, ExxonMobil's climate change denial funding totaled at least $33 million during the period 1997-2016.[32] "At least $33 million," because much of the funding has been channeled through dark identity-scrubbing groups such as Donors Trust and Donors Capital.

Even while facing multiple state investigations connected to its funding of climate change denial, Exxon was a top sponsor at the American Legislative Exchange Council (ALEC) annual meeting in July 2016, which discussed blocking Obama's Clean Power Plan. On its corporate board, Exxon has funded ALEC since 1981, and to the tune of at least $1,730,200 between 1998 and 2014.[33]

In 2016 disclosures, one of Exxon's larger single donations – $325,000 – went to the American Enterprise Institute, a critic of renewable energy that has long fought regulations to cut greenhouse gases. Its energy and environment "expert," Benjamin Zycher (BA, political science; PhD, economics), told the UN Paris climate summit in 2015 that "recent data" does not support greenhouse gases as a cause of polar sea ice collapse.[34] This was clearly deceitful: the heat-trapping nature of CO2 has been known since the mid-19th century.[35]

Secret Funding by Coal Companies: In April 2017, Peabody Energy, the country's largest investor-owned coal company, declared bankruptcy, following Arch Coal and Alpha Natural Resources. In all three cases, court-ordered disclosures revealed creditors well known as climate science deniers. These included Chris Horner, who regularly disparages climate science on *Fox* News and has called for investigations of IPCC and NASA scientists.[36]

According to the *Wall Street Journal*, Alpha paid Horner $18,600 before it declared bankruptcy. Meanwhile, the Free Market Environmental Law Clinic – an Alpha creditor – had paid him $110,000 in 2014, $115,865 in 2013, and $60,449 in 2012, according to the clinic's tax filings.[37]

Climate change denialism remains rampant. In July 2016, nineteen US Senators led by Sheldon Whitehouse called attention to the "web of denial," a network of think tanks and front groups that had executed "a colossal political scheme" to deceive the public about climate change, and to halt climate action.

As Dr. James Hansen had observed in 2012, this is "not an accident. There is a very concerted effort by people who would prefer to see business continue as usual."

Whitehouse was one of the first in Congress to propose a civil case, similar to the racketeering suit Bill Clinton brought against the tobacco industry, against fossil-fuel companies for deliberately misleading the public on climate science.[38]

Dr. Michael Mann sums it up: "The gulf between scientific opinion and public opinion has been bought with hundreds of millions of dollars of special interest money...The number of lives that will be lost because of the damaging impacts of climate change is in the hundreds of millions – to me, it's not just a crime against humanity; it's a crime against the planet."[39]

Climate Change Denial as a Crime Against Humanity

As cited earlier, a crime against humanity is "a deliberate act, typically as part of a systematic campaign that causes human suffering or death on a large scale."

We have established that the decades-long blocking and lying about scientific evidence on the dangers of human-caused global warming has been deliberate.

So the question arises, how many people have been, or will be, hurt or killed by climate change?

Many studies have been done over time. To cite a few:

- "Climate change is increasing the global burden of disease and in the year 2000 was responsible for more than 150,000 deaths worldwide. Of this disease burden, 88% fell upon children."[40]

- According to a March 2017 report from the Medical Society Consortium on Climate and Health, "a quarter of Americans can name one way in which climate change is affecting their health. This is seen by physicians across the country."[41]

- A 15-author 2016 report from the U.S. Global Change Research Program warns that people suffering chronic diseases such as Alzheimer's, asthma, chronic obstructive pulmonary disease, diabetes, cardiovascular disease, mental illness, and obesity are being threatened by climate change.[42]

A global estimate was supplied by an independent report[43] commissioned by 20 countries in 2012 to study the human and economic costs of climate change. The DARA study wrote that it:

> linked 400,000 deaths worldwide to climate change each year, projecting deaths to increase to over 600,000 per year by 2030…Heat waves kill many, to be sure, but global warming also devastates food security, nutrition, and water safety. Since mosquitoes and other pests thrive in hot, humid weather, scientists expect diseases like

malaria and dengue fever to rise. Floods threaten to contaminate drinking water with bacteria and pollution.

When the report looked at the added health consequences from burning fossil fuels – aside from climate change – the number of deaths jumps from 400,000 to almost 5 million per year. Carbon-intensive economies see deaths linked to outdoor air pollution, indoor smoke from poor ventilation, occupational hazards, and skin cancer.[44]

When disinformation known to be false is systematically used to deny dangerous realities that harm public health and kill millions of people, the deception clearly crosses the line to become a crime against humanity.

Conclusion

The 2014 IPCC 5th assessment Summary for Policy Makers, along with previous IPCC assessments, is solid proof of the unprecedented crime represented by today's level and rate of increase in atmospheric greenhouse gas pollution. It is definite because policy makers representing all world governments sit on the IPCC Panel and before the assessment can be published they scrutinize the assessment line-by-line for government approval.

As governments from high-emitting countries continue – against the will of their own citizens and of the nations most vulnerable to climate change – to allow the global climate catastrophe to unfold, they simply cannot say that they did not know. Participation in formulating the IPCC summaries makes the large GHG-polluting national governments undeniably culpable for their continued lack of action to bring about a rapid decline in global emissions.

Not only have they betrayed the IPCC science. While doing so they have pampered the lucrative fossil fuel industry with trillions of dollars in subsidies worldwide.

But worst of all they have failed to protect their citizens – now and for future generations.

This is the crime of all time.

Endnotes

1   Intergovernmental Panel on Climate Change, Working Group I, "Climate Change: The IPCC Scientific Assessment," Cambridge University Press, 1990. Working Group (WG) 1. *Summary for Policy Makers* (SPM), under "Executive Summary," page xi.
(http://www.ipcc.ch/ipccreports/far/wg_I/ipcc_far_wg_I_spm.pdf).

2   Ibid., page xxii.

3   Ibid., "Table 1: Summary of Key Greenhouse Gases Affected by Human Activities," p. xvi.

4   T.J. Blasing, CDIAC "Recent Greenhouse Gas Concentrations." Updated April 2016, Carbon Dioxide Information Analysis Center (http://cdiac.esd.ornl.govpns/current_ghg.html).

5   National Research Council, Board on Atmospheric Sciences and Climate, *Climate Stabilization Targets: Emissions, Concentrations, and Impacts over Decades to Millennia*, Section on "Stabilization Requires Deep Emissions Reductions," National Academies Press, 2011, 9.

6   Only Zero Carbon (http://www.onlyzerocarbon.org/ipcc_ar5.html).

7   United Nations Framework Convention on Climate Change, "Fact sheet: Climate change science - the status of climate change science today," February 2011(https://unfccc.int/files/press/backgrounders/application/pdf/press_factsh_science.pdf).

8   Yongyang Cai, Timothy M. Lenton, and Thomas S. Lontzek, "Risk of multiple interacting tipping points should encourage rapid CO2 emission reduction," *Nature Climate Change 6*, 21 March 2016.

9   Climate Interactive, "Climate Scoreboard: Where will the proposals from the climate negotiations lead?" © 2017 (https://www.climateinteractive.orgprograms/scoreboard).

10   Christiana Figueres, et al., "Three years to safeguard our planet," *Nature*, 28 June 2017.

11   Yongyang Cai, et al, "Risk of multiple interacting tipping points should encourage rapid CO2 emission reduction."

12   Silberg, Bog, "Why a half-degree temperature rise is a big deal," NASA, 29 June 2016 (https://climate.nasa.gov/news/2458/why-a-half-degree-temperature-rise-is-a-big-deal).

13   The IPCC author teams use a consensus process that is "based as far as possible on peer-reviewed and internationally available literature." This is explained more fully at IPCC, "The role of the IPCC and key elements of the IPCC assessment process," Geneva, 4 February 2010.

14   Naomi Oreskes and Erik M. Conway, *Merchants of Doubt: How a Handful of Scientists Obscured the Truth on Issues from Tobacco Smoke to Global Warming*, Bloomsbury Press, 2010.

15   "Smoking and Health Proposal," 1969. Original document from the Brown

& Williamson Records collection (https://www.industrydocumentslibrary. ucsfedu/tobacco/docs/#id=psdw0147).

16   Joe Walker, "Draft global climate science communications plan. Memorandum to API's Global Climate Science Team," 1998 (https://web.archive.org/ web/20070224102903/http://www.euronet.nl/users/e_wesker/ew@shell/ APIprop.html).

17   Denise Robbins, "How The Merchants Of Doubt Push Climate Denial On Your Television Blog," *Media Matters*, 8 March 2015 (https://www.mediamattersorg/ networks-and-outlets/heartland-institute?page=1).

18   Kathy Mulvey, et al., "The Climate Accountability Scorecard: Ranking Major Fossil Fuel Companies on Climate Deception, Disclosure, and Action," Union of Concerned Scientists, October 2016.

19   Katie Worth, "Climate Change Skeptic Group Seeks to Influence 200,000 Teachers," *Frontline*, March 28, 2017.

20   Ibid.

21   Ibid.

22   Alan Ehrenhalt, "'Dark Money,' by Jane Mayer," *The New York Times*, 19 January 2016.

23   Ben Jervey, "The Doubt Machine: Documentary Explores Koch Brothers' War on Climate Science," *The Real News*, 3 November 2016.

24   Ibid.

25   According to Moyers and Company, "At the very heart of the [Koch] network is Americans for Prosperity, a nationwide federated organization that now has paid staff in 34 states and contact lists for millions of conservative activists nationwide. AFP leverages its large financial war chest and grassroots contacts to oppose Democrats, help to elect very conservative Republicans and conduct policy campaigns to push those lawmakers in Congress and the states to enact Koch-supported policies and block or dismantle policies the network opposes." Alexander Hertel-Fernandez and Theda Skocpol, "Five Myths About the Koch Brothers – And Why It Matters To Set Them Straight," 1 March 2016.

26   Jason M. Breslow, "Bob Inglis: Climate Change and the Republican Party," *Frontline*, 23 October 2012.

27   Alexander C. Kaufman, "Don't Be Fooled By The Gentler Tone Of Charles Koch's Climate-Change Denial," *Huffington Post*, 26 June 2017, see video at 1:22 minutes.

28   Michael E. Mann, "I'm a scientist who has gotten death threats. I fear what may happen under Trump," *Washington Post*, 16 December 2016.

29   "'Shakedown' lawmaker is oil's darling in US House," Reuters, 17 June 2010 (http://af.reuters.com/article/energyOilNews/idAFN1724393720100617).

30   Suzanne Goldenberg, "Work of prominent climate change denier was funded by energy industry," *The Guardian*, 21 February 2015.

31   Neela Bannerjee, et al., "Exxon: The Road Not Taken," *Inside Climate News*, 15 September 2015.

32   "ExxonMobil's $33 Million Campaign to Sow Doubt and Denial about Global Warming," *DeSmogBlog*, n.d.

33   Nick Surgey, "ExxonMobil Top Sponsor at ALEC Annual Meeting," Center for Media and Democracy, 27 July 2016 (http://www.exposedbycmdorg/2016/07/27/ exxonmobil-top-sponsor-alec-annual-meeting).

34   Graham Readfearn, "ExxonMobil: New Disclosures Show Oil Giant Still Funding Climate Science Denial Groups," DeSmogBlog, 8 July 2016.

35   NASA, "Climate change: How do we know?" (https://climate.nasa. govevidence/).

36   Elliott Negin, "Coal Companies' Secret Funding of Climate Science Denial Exposed," *Huffington Post,* 13 April 2017.

37   Ibid.

38   Carpenter, Zoë, "Senator Sheldon Whitehouse on the Fossil-Fuel Industry's 'Web of Denial,'" *The Nation*, 13 July 2016.

39   Jervey, "The Doubt Machine," at 28:15 minutes. "The Doubt Machine: Inside the Koch Brothers' War on Climate Science," (https://vimeo.com/189639657) at 28:19

40   Perry E. Sheffield and Philip J. Landrigan, "Global Climate Change and Children's Health: Threats and Strategies for Prevention," *Environ Health Perspectives 119*, 291-298, 2011.

41   Kavya Balaraman, "Doctors Warn Climate Change Threatens Public Health," *Scientific American*, 17 March 2017. The full report is available at https:// medsocietiesforclimatehealth.org/reports/medical-alert/

42   A.J. Crimmins et al., eds., "The Impacts of Climate Change on Human Health in the United States: A Scientific Assessment," U.S. Global Change Research Program, Washington, DC, 312 pp. (http://dx.doi.org/10.7930/J0R49NQX).

43   DARA Internacional, *Climate Vulnerability Monitor: A Guide to the Cold Calculus of a Hot Planet, 2nd Ed.*, 2012 (http://daraint.org/climate-vulnerability-monitor/climate-vulnerability-monitor-2012/report/).

44   Rebecca Keberm, "Obama Is Right: Climate Change Kills More People than Terrorism," *New Republic*, 11 February 2015. The DARA Report is confirmed by the 2009 *Human Impact Report* from the Global Humanitarian Forum, which estimated 500,000 annual deaths from climate change, p. 13 (http://www.ghf-ge.org/human-impact-report.pdf).

Chapter Three

STATE CRIME AGAINST THE GLOBAL PUBLIC TRUST

Introduction

The duty of governments to protect the natural environment and our life-sustaining planet goes back to ancient Rome and is still extant.

In our present case the offense is global atmospheric pollution causing global climate and ocean disruption, which is a clear human rights violation against billions of vulnerable people and all future generations.

Regulation and banning of hazardous global pollution is well established over many decades. For example, cumulative environmental organic toxins such as persistent organic pollutants (POPs) have been banned internationally by the Stockholm Convention.

This is highly significant because anthropogenic CO_2 emissions are also a highly persistent and cumulative pollutant, taking thousands of years to be completely cleared from the atmosphere.

In 2007, the U.S. Supreme Court ordered the EPA to undertake an exhaustive scientific and public investigation based on the EPA's regulation of air pollutants. The 2009 EPA endangerment finding concluded that atmospheric GHG emissions are pollutants because they "endanger both the public health and the public welfare of current and future generations."[1]

But in spite of the 2007 Supreme Court order, "precisely nothing was done" by Washington over the next five years."[2] The Obama Administration failed to pass cap-and-trade legislation to control emissions in the two-year period they controlled Congress from 2009 to November 2010, and the issue received very little attention on the 2012 campaign trail.

In 2012, the EPA introduced regulations for GHG emissions from new passenger cars and trucks, and in 2015 produced standards under the Clean Air Act to limit carbon emissions from new coal and natural gas-fired power plants. These were an extremely modest start to the great totality of emissions that must be stopped.

The early history of the interpretation of the Clean Air Act and its application to reducing GHG emissions was a long and painful one, and it continues today. In the words of Brookings fellow Philip Wallach in 2012:

> Arguably, the most important effect of adopting sensible American climate change policy would be to put the United States in a position of global leadership on the issue. But the current CAA [Clean Air Act] policies, which are bitterly contested by half of our political class, are incapable of fulfilling this role...When critics say that our GHG emission controls are wasteful, poorly designed, and imposed by "fiat," they will be more than a little correct. Executing a legislative tradeoff enacting efficient carbon pricing and ending the strange interpretive odyssey under the Clean Air Act should be a priority for believers in sensible policy.[3]

The US government's stunning failure of political will to take sensible and sufficient action to curtail its deadly continuing CO2 emissions should be addressed as a state crime.

GHG emissions should also be regarded as an international

pollution crime under The UN Convention on Long-Range Transboundary Air Pollution, which entered into force in 1983. The greenhouse gas emissions of the high-emitting countries are already causing severe harm and deaths in the least-emitting countries.

In October 2014, the UN Office of Human Rights sent an open letter to the State Parties of the UN Framework Convention on Climate Change on the occasion of their meeting in Bonn, reminding them that:

> The responsibilities of the State Parties in all of the above respects should not be viewed as stopping at their borders. Climate change is a global threat to human rights that requires global cooperation to solve, in accordance with the principle of international cooperation firmly entrenched in the UN Charter, the Universal Declaration of Human Rights, and a host of human rights treaties and declarations. States must work together in good faith to protect the environment that supports and enables the full enjoyment of our human rights.[4]

In March 2014, Dr. James Hansen wrote from Beijing:

> Secretary of State John Kerry has offered to keep China informed of what we are doing about climate in the United States. If that is the best we can do, if we do not help China obtain the abundant, affordable carbon-free energy needed to raise living standards while leaving room on the planet for other species, I believe that our own children, and the world as a whole, are likely to look back on us as having been guilty of the world's greatest crime against humanity and nature.[5]

It is self-evident that atmospheric GHG pollution is the worst of all crimes, however crime is defined. The strongest legal case against greenhouse gas pollution crime is the Public Trust Doctrine, which is being pursued by Our Children's Trust and will be explored in Chapter 8.

Evidence of State Crime

In criminology, state crime is action or failure to act that breaks the state's own criminal law or public international law.

In Chapter 5 we will argue that the financing and production of fossil fuel energy, and the devastation they cause, violate three human rights Declarations of the United Nations, to the extent that they qualify as crimes against humanity. We will also see that the International Criminal Court has widened its scope to include environmental destruction.

The United Nations brooks no tolerance: "The prohibition of crimes against humanity, similar to the prohibition of genocide, has been considered a peremptory norm of international law, from which no derogation is permitted and which is applicable to all States."[6]

Article 5 of the Rome Statute of the International Criminal Court lists genocide, crimes against humanity, war crimes, and the crime of aggression as "Crimes Against Peace."

Unmistakably, global warming-induced water shortage and drought in Syria have played a direct role in the long conflict.[7] Professor Michael Klare's article "Climate Change as Genocide" links the 2017 famines in war-torn Nigeria, Somalia, South Sudan and Yemen to prolonged drought produced by climate change inaction, concluding that "inaction equals annihilation."[8]

Massive fossil fuel subsidies combined with the failure to reduce CO_2 emissions have promoted climate disruption worldwide – disruption that destroys peace through water-related violence. These government policies therefore qualify for consideration as crimes against humanity – and ultimately as genocide (if emissions are not urgently and emphatically reduced to avoid catastrophic tipping points).

In 2010, British barrister Polly Higgins proposed that the Rome Statute be amended to include ecocide as the fifth international Crime Against Peace.[9]

Ecocide is the destruction of the natural environment. Polly Higgins writes on her website:

> Every crime is defined by its elements – what lawyers call *Mens Rea* and *Actus Reus* (the state of mind and the act). Since the Paris Agreement, we have a situation where the *Mens Rea* in ecocide crime is that of recklessness, i.e., disregarding available information and proceeding/or failing to prevent, regardless of knowledge…no decision-maker can now claim ignorance of the potential harm resulting from dangerous industrial activity or failure to prevent it.[10]

Higgins' claim of recklessness in failing to reduce GHG emissions is incontestable. The preamble to the Paris agreement (signed by 195 countries) states this urgent imperative:

> *Recognizing* that climate change represents an urgent and potentially irreversible threat to human societies and the planet and thus requires the widest possible cooperation by all countries, and their participation in an effective and appropriate international response, with a view to accelerating the reduction of global greenhouse gas emissions,
>
> *Also recognizing* that deep reduction in global emissions will be required in order to achieve the ultimate objective of the Convention and emphasizing the need for urgency in addressing climate change…"[11]

The signed (and binding) Paris Agreement contained Article 2.1, which refers to the *objective* of the foundational UN Framework Convention of 1992:

This Agreement, *in enhancing the implementation of the [1992] Convention, including its objective*, aims to strengthen the global response to the threat of climate change, in the context of sustainable development and efforts to eradicate poverty, including by: (a) Holding the increase in the global average temperature to well below 2 °C above pre-industrial levels and to pursue efforts to limit the temperature increase to 1.5 °C above pre-industrial levels, recognizing that this would significantly reduce the risks and impacts of climate change.[12] [Emphasis added.]

The signed Paris agreement is therefore binding on the UNFCCC's *ultimate objective*, agreed to as foundational in 1992, which is:

Article 2 Objective

The ultimate objective of this Convention and any related legal instruments that the Conference of the Parties may adopt is to achieve, in accordance with the relevant provisions of the Convention, *stabilization of greenhouse gas concentrations in the atmosphere at a level that would prevent dangerous anthropogenic interference with the climate system.* Such a level should be achieved within a time frame sufficient to allow ecosystems to adapt naturally to climate change, *to ensure that food production is not threatened* and to enable economic development to proceed in a sustainable manner.[13] [Emphasis added]

This international consensus confirms global warming as "an urgent and potentially irreversible threat to human societies and the planet," thus creating a duty to act.

The Duty to Act

Beyond nuclear war, there is no form of environmental crime that can create a wider range of victims than unremitting greenhouse gas emissions.

The duty of governments to act to substantially reduce GHG emissions rests on at least three legal traditions:

1. Dating back to early English common law, the first duty of the king – and later of governments in Britain and America – was the obligation to protect the life, liberty and property of citizens.[14]
2. The idea that governments, by virtue of their sovereignty, have the responsibility to protect natural resources has a long history in the West, and is derived from the Roman Law of common properties. In the 6th century A.D., Justinian law stated: "By the law of nature these things are common to all mankind — the air, running water, the sea, and consequently the shore of the sea."[15]
3. The land and all of its resources belong to the people, especially in democracies, where sovereignty rests with the people. The public trust doctrine (PTD) says that government holds vital natural resources in trust for the people *in perpetuity.*

A stable, sustaining climate is required to ensure the security of these vital natural resources, which are increasingly held to include the atmosphere (as will be discussed in Chapter 8 under "atmospheric trust legislation"). Thus there is a duty for government to regulate GHG emissions to secure a stable climate *for posterity.*

When the known impacts of unreduced GHG emissions – extreme weather events, sea level rise, food insecurity, epidemic diseases, and the displacement of millions of people (impacts that are already unfolding) – are not met head-on by governments, the abject failure to regulate or mitigate these emissions is both a state crime and an international crime that betrays the public trust.

State Blocking of Climate Action

During the last ten UN climate conferences, the large GHG-polluting national governments not only committed the crime of omission by failing to protect their citizens from climate disruption: they blocked and delayed action needed to save vulnerable non-polluting nations from CO_2-induced havoc already underway. The United States often acted as chief procrastinator:

- In 2007, more than 200 leading climate scientists petitioned government leaders at the Bali conference to take radical action. At the insistence of the Bush administration there was no mention of specific CO_2 reduction targets. Delegate Kevin Conrad from Papua, New Guinea told the Americans: "We seek your leadership. But if for some reason you are not willing to lead, leave it to the rest of us. Please get out of the way."[16]
- In a book titled *State Crime and Resistance*, the authors write: "The Copenhagen Conference in 2009 may have been the last, best opportunity for the governments of the world to act forcefully to prevent catastrophic climate change. Yet the administration of Barack Obama and the entire international political community utterly failed to take any strong actions that might avert the impending ecocide. As John Sauven, Executive Director of Greenpeace UK, bluntly stated after the failure of this conference, 'The City of Copenhagen is a crime scene tonight, with the guilty men and women fleeing to the airport. There are no targets for carbon cuts and no agreement on a legally binding treaty.'"[17]

- "Our collective failure to take action against global warming," Mark Hertsgaard pointed out in 2011, "had been a conscious decision, a result of countless official debates where the case for reducing greenhouse gas emissions was exhaustively considered and deliberately rejected."[18]
- Vandana Shiva said of the 2012 Rio+20 conference: "The entire energy of the official process was focused on how to avoid any commitment. Rio+20 will be remembered for what it failed to do during a period of severe and multiple crises and not for what it achieved."[19]
- In 2012, a summit of nearly 200 nations in Doha, Qatar succeeded in keeping the Kyoto Protocol alive, but just barely, because Canada, Russia, New Zealand, and Japan had opted out (the US had not ratified Kyoto in the first place), with the remaining countries accounting for only 15% of global greenhouse gas emissions.
- In 2014, top scientists at the IPCC were "at their wits' end on how to make world leaders realize that the Earth is on the brink of a climate catastrophe if they do not cooperate in cutting down their gas emissions before mid-century." However, there was "still time for these countries to avert the impending catastrophe if only they will aggressively decrease the concentration of greenhouse gases."[20]
- Weeks before Donald Trump abandoned the Paris accord, Dana Nuccitelli wrote: "The [Trump] administration has effectively declared war on the Earth's climate and our future well-being. Noam Chomsky has called the Republican Party the most dangerous organization in human history because of its climate denial and policy obstruction."[21]

Bribery and Corruption in Climate Politics

Transparency International's *Corruption Perceptions Index* shows that many developing countries suffer from pervasive systemic corruption that siphons off billions of dollars in revenue, undermining basic services while benefitting the few.[22]

Problems are also caused in higher-ranked countries such as the US, Canada, the UK and Europe, not only as the source of bribery and corruption, but by legislation that circumvents anti-corruption laws.

A striking US example was the subversion of the 1907 Tillman Act, which prohibited direct political campaign contributions by corporations.

In January 2010, the Supreme Court ruled in its *Citizens United v. Federal Election Commission* decision that corporations are persons with the right to free speech. This includes the right to spend unlimited communications funds to influence election results, as long as it is not spent through direct contributions to candidates. During the 2010 midterms the corporate floodgates opened and outside groups spent more campaign money than they had in all the midterms combined since 1990. The Republicans took control of Congress, and voted nearly 200 times to block and delay Obama's environmental legislation.

In 2013, the year that atmospheric CO_2 reached 400 ppm for the first time in over 2.5 million years, 163 elected representatives from the 113th Congress had "taken over $58.8 million from the fossil fuel industry that is the driving force behind the carbon emissions that cause climate change … And their constituents are paying the price, with Americans across the nation suffering 430 climate-related national disaster declarations since 2011."[23] As of 2013, more than 56 percent of the Republican members of the Congress denied climate science, with 90 percent of the leadership positions filled by climate science deniers.[24]

In June 2017 – 110 years after political mega-bribery had been outlawed by the Tillman Act, and while American concern about global warming was at a 3-decade high – the Republican-backed fossil fuel industry got what it wanted:

> The big-money supporters got a return on their investment last week, when 22 Republican senators whose campaigns have collected more than $10m in oil, gas and coal money since 2012 sent a letter to the president urging him to withdraw from the Paris deal.[25]

As Oil Change International put it:

> The fossil fuel industry undoubtedly has a stranglehold on U.S. democracy, bribing elected officials with campaign contributions and pressuring them with millions of dollars of spending in strong-armed lobbying. In return, the industry is provided massive subsidies while raking in mind-boggling profits and giving their executives lavish compensation packages.[26]

Other governments accept industry bribes as well. In 2017, leaked emails showed that top Shell executives knew that its $1.1 billion bribery paid to the Nigerian government for one of Africa's most valuable oil blocks would go to a convicted money-launderer and not to the Nigerian people.[27] The BBC reported that the bribe was "equivalent to more than the 2016 ministry of health budget in Nigeria – a country where one in ten children don't live to see their fifth birthday."[28]

These flagrant state crimes require sustained and widespread civil resistance, which will be discussed in Chapter 11.

State-Corporate Crime

State-corporate crime |refers to "serious social harms that result from the interaction of political and economic organizations."[29] Professor of Sociology Ronald Kramer has advanced the idea that the crime of omission to regulate GHG emissions can best be conceived as state-corporate criminality.[30]

Professor of Criminology Rob White also stressed the importance of investigating the state as a major facilitator of harm, either on its own or in conjunction with special interests.[31]

In 2011, Prof. White described the failed Copenhagen conference as a state-corporate crime: "The abject failure of the Copenhagen talks to actually do something about carbon emissions and to address climate change issues in a substantive fashion is a striking example of the fusion of state and corporate interests to the detriment of the majority."[32]

British sociologist Anthony Gidden defined the state's vital role, writing that "the state must be the prime actor" in addressing climate change."[33] Kramer suggests that the state's failure to do so constitutes "negligent state criminality," causing victimization and unnecessary loss of life.[34] Nowhere is the state more deeply engaged in climate crime than in fossil fuel subsidies.

Fossil Fuel Subsidies as a Violation of Democracy

Polls have shown that about 70% of Americans want their governments to stop subsidizing fossil fuels.[35] Nearly 90% polled the same sentiment in 2016.[36]

In 2017, 73% of Americans prioritized alternative energy over oil.[37]

Added to this overwhelming civic opposition, subsidizing a long-standing industry is contrary to free market economics:

> In the United States, fiscal conservatives are militant about letting free markets and market forces make our energy choices rather than allowing government policies to "pick winners." But in Congress, the same fiscal conservatives are silent about government subsidies for coal, oil and gas.[38]

US Congressional silence is not unique. By 2017, G20 governments who signed the Paris agreement were still giving preferential treatment to the outmoded oil industry, providing four times more public finance to fossil fuels than to clean energy.[39]

In Australia, fossil fuel subsidies were likened to "being in bed with big tobacco." "Continuing to fund polluters when we know the damage being done to the environment is unforgivable intergenerational theft," said Luke Stickels of the Australian Education Union, while he and 50 civil society groups protested Australia's $7.7 billion in subsidies.[40]

Why is energy financial policy so irrational? Alden Meyer of the Union of Concerned Scientists explains:

> The main barrier to confronting the climate crisis isn't a lack of knowledge about the problem, nor is it the lack of cost-effective solutions. It's the lack of political will by most world leaders to confront the special interests that have worked long and hard to block the path to a sustainable low-carbon future.[41]

These world leaders have a *primary democratic duty to act* – to legislate the clear priority of their citizens for clean energy, and to protect the public from greed-driven corporate profiteering.

Environmental Racism

Another human rights focus to global warming is its disproportionately negative impact on communities and nations of colour, and on low-income communities.

Many of these communities are economically unprepared to adapt to climate change impacts. Included in this group are indigenous peoples closely tied to the land and unable to adapt to impacts such as sea-level rise that affect their livelihood and cultures.

Also included are the world's born and unborn children, most of whom cannot fight vested interests for their natural right to a stable future climate.

Those unable least able to protect themselves from state-corporate complicity in promoting fossil fuels are shouldering far more environmental burdens than the rest of society, both nationally and globally. This too is criminal climate injustice. GHG emissions must stop.

Endnotes

1   United States. Environmental Protection Agency. "EPA Finds Greenhouse Gases Pose Threat to Public Health, Welfare / Proposed Finding Comes in Response to 2007 Supreme Court Ruling," 17 April 2009.

2   Philip A. Wallach, "U.S. Regulation of Greenhouse Gas Emissions, *Governance Studies at Brookings*, 26 October 2012, 3-4.

3   Ibid., 13.

4   United Nations Human Rights. Office of the High Commissioner. "Climate change is a global threat to human rights, UN experts warn States involved in climate negotiations," 17 October 2014.

5   James E. Hansen, "World's Greatest Crime against Humanity and Nature," 10 March 2014 (http://www.columbia.edu/~jeh1/mailings/2014/20140310_ChinaOpEd.pdf).

6   United Nations. Office on Genocide Prevention and the Responsibility to Protect, "Crimes Against Humanity: Definition" (http://www.un.org/en/genocideprevention/crimes-against-humanity.html).

7   Peter H. Gleck, "Water, Drought, Climate Change, and Conflict in Syria," *American Meteorological Society Online Journal*, 3 February 2014.

8   Michael T. Klare, "Climate Change as Genocide: Why Inaction Equals Annihilation," *Common Dreams*, 20 April 2017.

9   Eradicating Ecocide, "What is Ecocide? Proposed Amendment to the Rome Statute," n.d. (http://eradicatingecocide.com/the-law/what-is-ecocide).

10   Eradicating Ecocide, "Why Ecocide Crimes are Crimes of Recklessness," 27 February 2017 (http://eradicatingecocide.com/2017/02/27/why-ecocide-crimes-are-crimes-of-recklessness).

11   UNFCCC, "Adoption of the Paris Agreement," 12 December 2015, 1 (https://unfccc.int/resource/docs/2015/cop21/eng/l09r01.pdf).

12   Ibid., 22.

13   United Nations Framework Convention on Climate Change FCCC Convention Text, signed and ratified in 1993 (http://www.globelaw.com/Climate/fcc.htm).

14   Steven J. Heyman, "The First Duty of Government: Protection, Liberty and the Fourteenth Amendment," n.d. (http://scholarship.law.duke.edu/cgi/viewcontent.cgi?article=3172&context=dlj).

15   *The Institutes of Justinian* (535 A.D.).

16   "U.S. backs down after needling," *Toronto Star*, 16 December 2007.

17   "Public Criminology and the Responsibility to Speak in the Prophetic Voice Concerning Global Warming." Pp. 41-53 in Elizabeth Stanley and Jude. McCulloch (Eds.), State Crime and Resistance. London: Routledge, 41-53..

18   Mark Hertsgaard, *Hot: Living Through the Next Fifty Years on Earth*, Houghton Mifflin Harcourt, 2011, 12.

19   Vandana Shiva, 2012. "Rio+20: An undesirable U-turn," *The Asian Age*, 25 June 2012, 1.

20   Desiree Q. Sison, "Scientists Warn of Catastrophic Climate Change by 2050," *China Topix*, 14 April 2014.

21   Dana Nuccitelli, "NY Times' Stephens can't see the elephant in the room on climate change," *The Guardian*, 16 May 2017.

22   Transparency International, *Corruption Perceptions Index, 2016* (https://www.transparency.org/news/feature/corruption_perceptions_index_2016).

23   Tiffany Germain et al., "The Anti-Science Climate Denier Caucus: 113th Congress Edition," *Climate Progress*, 26 June 2013.

24   Ibid.

25   Tom McCarthy, "The Republicans who urged Trump to pull out of Paris deal are big oil darlings," *The Guardian*, 1 June 2017.

26   Oil Change International, "The Price of Oil: Corruption" (http://priceofoil.org/thepriceofoil/corruption).

27   Global Witness, "Shell Knew," 10 April 2017 (https://www.globalwitness.org/en-gb/campaigns/oil-gas-and-mining/shell-knew/).

28   Simon Jack, "Shell corruption probe: New evidence on oil payments," *BBC News*, 10 April 2017.

29   Kramer, Ronald C. "Climate Change: A State-Corporate Crime Perspective." In: Toine Sappens et al., eds., *Environmental Crime and its Victims*, 2014, 29-39.

30   Kramer, "Climate Change."

31   Rob White, *Transnational Environmental Crime: Toward an Eco-global Criminology*, Routledge, 2011, 13.

32   Ibid., 148.

33   Anthony Glidden, *The Politics of Climate Change*, 2nd ed., 2011, Polity Press, 94.

34   Kramer, "Climate Change."

35   "Do Americans Support or Oppose Subsidies For Fossil Fuels?" Yale Program on Climate Change Communication, 14 February 2012 (http://climatecommunication.yale.edu/publications/do-americans-support-or-oppose-subsidies-for-fossil-fuels).

36   Michael Slezak, "Fossil fuel subsidies: majority of voters in mining states opposed tax breaks," *The Guardian*, 17 May 2016.

37   Zac Auter, "In U.S., 73% Now Prioritize Alternative Energy Over Oil, Gas;" Gallup, 24 March 2017 (http://www.gallup.com/poll/190268/prioritize-alternative-energy-oil-gas.aspx).

38   William S. Becker, "How Can We Pay for the New Energy Economy?" *Huffington Post*, 19 December 2016.

39   Alex Doukas, "Talk is Cheap: How G20 Governments are Financing Climate Disaster," Oil Change International, July 5, 2017 (http://priceofoil.org/2017/07/05/g20-financing-climate-disaster).

40   Thom Mitchell, "$7.7 Billion Fossil Fuel Subsidies 'Like Being In Bed With Big Tobacco,'" *New Matilda*, 26 April 2016.

41   Alden Meyer, "World Leaders Lack Political Will to Make Progress at Climate Negotiations in Bonn: Statement by Alden Meyer," Union of Concerned Scientists, 3 May 2013.

Chapter Four

MEDIA
COLLUSION

"The news must be comprehensive
and proportional to the significance of events."

Kovach and Rosenstiel,

The Elements of Journalism (2014)

During the three weeks of the 2016 United States presidential debates, cities in 34 states reported record high temperatures.[1]

Yet there was not a single question asked in the six hours of debate about climate change.

The first debate, on September 26, was the most watched presidential debate in U.S. history, attracting 84 million viewers. Moderated by Lester Holt of NBC, it was introduced by well-known NBC journalists Savannah Guthrie and Chuck Todd.

In the background, and not mentioned by these household names, hung the small sign, "Commission on Presidential Debates".[2]

A commission is generally understood to be "a group of people officially charged with a particular function" (Oxford), or "a government agency having administrative, legislative, or judicial powers" (Merriam Webster).

This "commission" is a non-profit organization that was founded by, and has been jointly sponsored by, the Democratic and Republican parties since 1987.

It is exempt from federal tax. Little is known about who funds it, but in a 2004 study, 93% of its contributions came from six anonymous corporate donors.[3] Although it claims to be non-partisan, at its inception the chairmen agreed that third-party candidates should be excluded from the debates, and rules were established to disqualify them.

The Commission on Presidential Debates (CPD) selects its moderators, and according to its website, "the moderators alone select the questions."

The public knows very little about the CPD; all people sense is that an official commission of some kind is conducting the presidential debates through well-known media personalities that they tend to trust.

A 2016 Gallup study showed that 64% of Americans were worried about global warming.[4] How are we to reconcile this poll with the fact that between them, the "commission" and its four moderators (representing NBC, CNN, Fox, and ABC News) failed to ask a single question about climate change during the six hours of debate?

Clearly the Commission on Presidential Debates has a misleading name. It is not a neutral public body that oversees fairness and objectivity in questioning presidential candidates. It, and the networks that carry it, have been politicized by fossil fuel contributions and advertising to champion business-as-usual – at the expense of public values.

In 2015, the year of the greatly anticipated Paris climate summit, an event of great moment occurred: the founding of *The Subnational Global Climate Leadership* (SGCL).

A total of 187 jurisdictions, representing 38 countries and six continents, have signed or endorsed the SGCL "Memorandum of Understanding" to limit warming by 2050 to below 2° Celsius. This coalition represents more than 1.2 billion people and $28.8 trillion in GDP – equivalent to 16% of the global population and 39% of the global economy.[5]

UN Secretary-General Ban Ki-moon declared the SGCL to be a game-changer.

What is worse than the politicized presidential debate omission is that during the two years since its inception, the SGCL *was not reported once* in the US or Canadian mainstream media (apart from a one-liner in *Time Magazine*) – including the Canadian public broadcaster, the CBC.

There is no benign explanation for a full media blackout of a significant global development that was heralded by the United Nations Secretary-General.

This blackout goes far beyond ignorance or negligence. It is a willful obstruction of public knowledge of the extraordinary extent of global efforts to combat the greatest existential threat of all time by changing business-as-usual.

We define this willful, methodical blocking of vital survival information as an unprecedented crime against life on the planet.

The First Purpose of Journalism

In their 2014 book, *"The Elements of Journalism: What Newspeople Should Know and the Public Should Expect,"* Bill Kovach and Tom Rosenstiel emphasize that the first purpose of journalism is

> to provide people with the information they need to be free and self-governing … Its first obligation is to the truth … its first loyalty is to citizens.[6]

and

> the purpose of news is to help people self-govern, but that only begins with giving people the information they need to do so. News must also be about solving the problems that confront individuals and the community.[7]

Returning to our example of the presidential debates, the topics were weighted as follows: six questions about the

Syrian civil war, four questions about terrorism, three questions each about US-Russia relations, immigration, job creation, Trump's taxes, and Clinton's emails, and single questions about "birtherism," cyberterrorism, Trump's twitter posts, and Clinton's "basket of deplorables."

Incredibly, there were no policy questions asked about the fundamental issues that directly impact the lives of the American people: health care, education, student loans, housing affordability, pensions, or labor unions.

Instead, 84 million people were duped by the networks into a charade that echoed the media's steady war propaganda drumbeat: Syria, terrorism, and Russia.

Peter C. Goldmark, Chairman and CEO of the *International Herald Tribune,* explains that corporate media leadership has a "solemn fiduciary responsibility arising from their ownership of a news organization – that they hold a public trust."[8]

How have the U.S. television networks honored this public trust?

Ross Gelbspan, a Pulitzer-prize winning journalist, reported back in 2004:

> A few years ago I asked a top editor at CNN why, given the increasing proportion of news budgets dedicated to extreme weather, they did not make [the connection to global warming]. The editor said, "We did that. Once. But it triggered a barrage of complaints from the Global Climate Coalition to our top executives at the network." (The Global Climate Coalition was, at the time, the main fossil fuel industry lobbying group opposing action on global warming.)[9]

This conversation apparently occurred in 1999.[10] The CNN network has therefore had nearly 20 years to provide the climate information people need to be self-governing in the matter. During the hottest autumn on record, it could easily have raised the issue in the 2016 presidential debates. But *when it counted so much*, in company with NBC, ABC, and Fox News, CNN chose not to question the two presidential candidates – one of whom would preside over US climate policy for the next four crucial years.

Following these debates the new administration appointed climate change denier Scott Pruitt to head the Environmental Protection Agency. Later, President Trump withdrew from the Paris Climate Accord, making a mockery of responsible U.S. leadership worldwide.

The Impact of Media Denial on Public Beliefs and Actions

International polls confirm that US newspapers are unique in promoting climate promoting climate denial.

A 2012 story, "American Newspapers Are Number One in Climate Denial," reported a study that compared *The New York Times* and *Wall Street Journal* with leading newspapers in Brazil, China, France, India, and the United Kingdom:

> America is unique when it comes to giving a platform to climate deniers and skeptics. According to a new analysis of data released [in 2011], American newspapers are far more likely to publish uncontested claims from climate deniers, many of whom challenge whether the planet is warming at all.[11]

Since 2012, U.S. media denial has gotten worse. In 2016, a team of sociologists at the University of Oklahoma found that Republicans are less likely today to accept the reality and effects of human-caused global warming than they were a decade ago. The study identified conservative media climate change denial as the root influence. The *Wall Street Journal*, for example, featured only 31 climate science articles out of 93 climate-related opinion pieces from January 2015 to August 2016.[12]

A 2015 study found that the *WSJ's* biased climate coverage extended beyond its opinion pages to its news coverage.[13] The following year its pages were blanketed with defenses of Exxon Mobil.[14] Since Rupert Murdoch (a good friend of Donald Trump) purchased the *WSJ* in 2007, the paper has bombarded its readers with climate disinformation and fossil fuel industry propaganda, usually without disclosing the authors' ties to fossil fuels.

Media Matters studied op-eds, editorials, and columns in *The New York Times*, the *Wall Street Journal*, *USA Today*, and the *Washington Post* from January 1, 2015, to August 31, 2016, to determine how often climate science denial appeared on each newspaper's opinion pages. Although all the editorial boards (except the *WSJ*) accept the climate science consensus, only the *New York Times* refrained from publishing climate science denial pieces.[15]

Reflecting US media denial, world polls in 2014 and 2016 showed that U.S. public concern is less than concern elsewhere.

A 2014 poll of 20 wealthy countries found that the American people lead the world in denial, with 52 percent of the population stating that climate change is a natural phenomenon, not CO2-related, and denying that we are headed for environmental disaster unless we change our habits quickly.[16]

The Pew Research Center reported in a 2016 global attitudes survey that "majorities in all 40 nations polled say climate change is a serious problem, and a global median of 54% believe it is a *very* serious problem." The immediacy of climate change (now or in the next few years) worried 95% of Latin Americans, 86% of Europeans, 85% of Africans, and 69% of Americans.[17]

It is thus evident that the US corporate media has favored its political connections and profits over its first responsibility to provide the information necessary for the American people to understand climate change and be self-governing in the matter. The same is true for Canada.

The Responsibility of the Media to Help Solve Problems

Kovach and Rosenstiel, knowing that the public no longer trusts the press, have for years been trying to right journalism's wrongs.

Given their view that a central purpose of the news is to help people self-govern, tthey explain that people need the news to act for them as "empowerer," by

> providing audiences with tools and information so they can act for themselves. This involves making information interactive, providing dates when action needs to be taken, explaining how to get more involved. It may go even further and involve organizing events that bring the community together to solve problems.[18]

In 2016, with respect to climate change impacts, the media dismally failed to fulfill this function in both Canada and Australia.

Fort McMurray, Alberta, is the epicenter of the enormous and controversial oil sands energy mega-project. In May 2016, starting with an unusually hot, dry air mass, record-setting temperatures, and high winds, an unprecedented firestorm raged right through to July, forcing more than 80,000 people to flee the city, where 2,400 homes were destroyed.

Four Canadian communications and journalism professors reported that when a few politicians, environmentalists, scientists, and journalists "cautiously raised the connection between the wildfire and climate change, many – including Canadian Prime Minister Justin Trudeau – not only refused to draw these linkages but also questioned those who suggested them."[19] Controversy raged in the national newspapers, many claiming that it was "insensitive" to raise climate change during the inferno.

In north-east Australia, the corals of Lizard Island of the Great Barrier Reef, which had recently been in "full glorious health," were described in 2016 by a returning diver as rotten and foul-smelling:

> It was one of the most disgusting sights I've ever seen. The hard corals were dead and covered in

algae, looking like they've been dead for years. The soft corals were still dying and the flesh of the animals was decomposing and dripping off the reef structure.[20]

Dozens of climate and coral reef scientists believe the widespread reef devastation to have been caused by globally-warmed oceans. Frustrated that Queensland's largest daily newspaper (the *Courier Mail,* owned by Rupert Murdoch) has been consistently failing to provide accurate coverage of coral bleaching, these scientists placed an ad in the *Courier* describing the magnitude of the problem and its connection to coal mining, export, and burning.

The Canadian professors concluded "that in both countries, the boundaries of public discourse about climate change are being constrained by powerful sections of the media and political establishments."[21]

Far from empowering people to solve problems, constraining public debate is a deliberate and therefore criminal impediment to human survival in the face of increasingly lethal firestorms, droughts, floods, and rotting ocean life.

The Responsibility of TV Meteorologists

For years weathercasters have repeated the mantra, "No single weather event can be blamed on climate change." One would assume that such a statement from a meteorologist would be based on an understanding of the climate science consensus.

However, a 2016 survey of 4,062 American Meteorological Society (AMS) members about climate change reveals that just 33% of respondents hold a bachelor's degree or higher in meteorology, and only 37% considered themselves to be expert in climate science.[22]

This may explain why only 29% agreed with the wording of the scientific consensus that climate change has been caused "largely or entirely by human activity." A further 38% believe that climate change is caused 60-80% by human activity. The combined figure, 67%, was close to the general U.S. population figure of 65%, and is a long way from the 97% consensus of published climate scientists.[23]

It's the blind leading the blind. Twenty percent of the AMS membership are TV meteorologists who have such doubt in their minds about the source of climate change that they cannot begin to assign attribution, or to approach exercising the media's responsibility to help citizens to deal with it.

A case in point: During the extreme eastern US. heat wave of Christmas 2015, only one out of thousands of TV weather forecasters linked the record temperatures to climate change. Some even went out of their way to explain why the temperatures could not be tied to climate change, one calling it "irresponsible" to do so.[24]

However, Dr. Steve Pacala of Princeton University explained in February 2017 that climate disasters have become so frequent that they could not occur statistically unless humans were influencing them. He said journalists are now able to attribute the cause and severity of climate events to our fossil fuel emissions. This will show citizens how much climate change is costing, countering the idea that addressing climate change is too expensive.[25]

With U.S. news reporting on climate change so highly politicized, it's not surprising that weathercasters have a vested interest in retaining their conservative viewers (and jobs). This is nowhere more apparent than in the behavior of the right-leaning Sinclair Broadcast Group, *the largest owner of TV stations in the United States.* As of May 2017, its 173 stations were affiliates of the ABC, CBS, NBC, and Fox news networks.

The Politicization of U.S. Climate News

For several years, Sinclair stations ended their news programs with a commentator speaking of the "manmade global warming hysteria that swept the nation," while claiming that "most people realize manmade global warming is a hoax."[26]

CNN is no better. On April 22, 2017, the day after the earth's atmosphere reached a terrifying CO_2 concentration of 410 ppm, and the week before a major climate march, CNN's *New Day Saturday* held a panel discussion.

The debate featured not climate scientists, but mechanical engineer Bill Nye the Science Guy, as the climate-change spokesperson, and atomic physicist William Happer, a climate change denier who has argued that the "demonization of CO_2" "really differs little from the Nazi persecution of the Jews, the Soviet extermination of class enemies or ISIL slaughter of infidels."[27]

Happer stated that CO_2 was not a harmful pollutant but a benefit to the planet, and called for the cancellation of the Paris climate agreement because it "doesn't make any scientific sense. It's just a silly thing."

The same day, *CBS Weekend News* also covered the marches, including a report on rapidly melting Arctic ice and the future impacts of climate change. But it also ran a segment, "Climate Realists," where it interviewed Joseph Bast, head of the climate-denying Heartland Institute.

Bast, who is not a scientist, falsely argued that climate change is simply the natural order of things and is beneficial because of decreased deaths from cold.[28]

The real failure here is not that CNN and CBS chose to air falsely "balanced" opinions. Or that they aired sensational, outrageous views to attract viewers. As Bill Nye said to his CNN host – reflecting the overwhelming percentage of scientists who believe that climate change is real – "you're doing a disservice by having one climate change skeptic and not 97 or 98 scientists or engineers concerned about climate change."[29]

Yet they well understood this disservice because the 2015 Paris summit made history when the whole world formally agreed to confront climate change. This reckless national network airing of extreme climate denial claims the very week CO_2 hit 410 ppm was negligent if not criminal.

False Climate Information as Media Crime

First, to define our terms:

> "Misinformation" is information that is false, but the person who is disseminating it believes it to be true. It is therefore only negligent.
>
> "Disinformation" is information that is false, and the person who is disseminating it knows it is false. It is a deliberate lie, and serves as propaganda.

We cannot always prove the intent of the media, but we do know that the scientific state of climate change has been announced by NASA and NOAA media advisories for decades.

Extraordinary, then, (and in spite of the *Media Matters* study above reporting *NYT's* non-climate-change-denial editorial board) that in April 2017 *The New York Times,* which had shut down its 9-person climate desk in 2013, hired neoconservative journalist Bret Stephens as an op-ed columnist. Stephens had backed the invasion of Iraq in 2003 and had dismissed climate change as an "imaginary enemy" during the 2015 Paris climate summit.

His first column, dated April 28 and titled "Climate of Complete Certainty," questioned the value of the certainty flowing from the scientific consensus on climate change.

His response to a firestorm of reader protest was that he's not a climate change denier, but merely a "climate agnostic," because the science remains unsettled and that more research is needed before acting decisively. What research, exactly, is missing? Is he qualified to say? His high-sounding "agnosticism" echoes Big Tobacco's manufacturing of doubt in the 1960s.

The enormously influential media outlets named above are ideally positioned to help people understand that global warming is not a far-off, distant threat, but is visibly happening all around them, right now, and is profoundly affecting their lives.

By failing to inform viewers of the full truth of climate change, TV stations and newspapers are guilty of far more than neglecting their social responsibility on a daily basis.

They are turning their backs on proven realities at the eleventh hour, 25 years after 172 nations attended the Rio Earth Summit, agreeing on the UN climate change convention.

This intentional choice goes beyond misinformation (negligence) and even disinformation (propaganda).

This choice goes beyond negligence:

> Failure to use the degree of care appropriate to the circumstances, resulting in an unintended injury to another.

This choice goes further, beyond culpable negligence:

> Negligent actions committed with a disregard of the consequences.[30]

The failure to use the degree of media care appropriate to preventing CO2 emissions causing ever-worsening floods, wildfires, sea level rise and mass migrations, fits the legal description of criminal negligence:

> The failure to use reasonable care to avoid consequences that threaten or harm the safety of the public and that are the foreseeable outcome of acting in a particular manner.[31]

In 2013, Tom Engelhardt, recognizing that genocide is usually considered the ultimate crime, coined the term "terracide" to describe an even more ultimate crime, writing:

> To destroy our planet with malice aforethought, with only the most immediate profits on the brain, with only your own comfort and wellbeing (and

those of your shareholders) in mind: Isn't that the ultimate crime? Isn't that terracide?[32]

The fossil-fuel companies are guilty of the ultimate crime, he said, because they are earning their "profits directly off melting the planet, knowing that their extremely profitable acts are destroying the very habitat, the very temperature range that for so long made life comfortable for humanity."[33]

However, Big Carbon could never have been able to continue its polluting ways – long after the scientific community had reached consensus about the connection between fossil-fuel emissions, global warming, and climate change – without the assistance of the media.

Corporate Media Coverage Declines as Warming Increases

Compounding the issue is that the overall media coverage of climate change has steadily declined since 2009.

Media Matters has kept track of the *combined annual minutes of climate coverage* by ABC, NBC, CBS, and Fox News networks since 2009, the year of the Copenhagen climate summit, when a total of 205 minutes were aired.

Then followed 2010 (48 min.); 2011 (47 min.); 2012 (67 min.); 2013 (129 min.); and 2014 (154 min.).[34]

Surprisingly, 2015, the year of the much anticipated Paris climate summit held in November-December, the figure dropped to 146. The summit was overshadowed for much of its two-week duration by the San Bernardino shootings in California and by Donald Trump's behavior, which dominated the headlines in 2015.

In 2016, the combined coverage dropped 66% to 50 minutes for the four networks, even though 2015 had been declared by NASA as by far the hottest year on record.

While in 2016 and 2017 the media began drum-beating the words "climate change," it did so from the point of view of *adapting* to it rather than *preventing* it. *A corporate media declaration that business-as-usual must change to prevent future catastrophe is never heard.*

Sources of Responsible Media

With the corporate newspapers and TV networks virtually a write-off for consistent, non-political, evidence-based climate information, there are, fortunately, many alternatives online.

The British daily newspaper, *The Guardian* (free online), has been around since 1821 and runs researched articles on the climate crisis daily.

The Nation (a free weekly online*)*, founded in 1865 and headquartered in Washington DC, is financed mostly by donors and reports regularly on climate change.

There are too many online progressive news sources to list, but to name a few, *Climate Progress, Yes! Magazine, Mother Jones, AlterNet, and TruthOut* all publish on climate change.

Peer-reviewed general science journals such as *Scientific American* and *Nature* are available by subscription.

ScienceDaily offers *Climate News* online; NASA offers regular articles on the website, "Global Climate Change: Vital Signs of the Planet," and the Union of Concerned Scientists runs the website, "Confronting the Realities of Climate Change."

Independent sources publishing without a vested interest in fossil fuels demonstrate the almost universal consensus that climate change is real and is causing global weather upheaval.

The world can only hope that the U.S. corporate media owners will soon abandon their self-interest and embrace the truth spoken by Tom Engelhardt in 2014:

> The future of all other stories, of the news and storytelling itself, rests on just how climate change manifests itself over the coming decades or even century. . . Climate change isn't the news and it isn't a set of news stories. It's the prospective end of all news.[35]

Endnotes

1   David Leonhardt, "The Debates Were a Failure of Journalism," *The New York Times*, 20 October 2016.
2   "The First Presidential Debate," Hillary Clinton And Donald Trump (Full Debate)," *NBC News*, 26 September 2016 (https://www.youtube.com/watch?v=855Am6ovK7s)
3   "Two-party debates: A Corporate-Funded, Party-Created Commission Decides Who Debates — and Who Stays Home," The Center for Public Integrity, 18 September 2008.
4   Gallup, "U.S. Concern About Global Warming at Eight-Year High," March 16 2016.
5   The Subnational Global Climate Leadership, Under2° (http://under2mou.org/).
6   Bill Kovach and Tom Rosenstiel, *The Elements of Journalism: What Newspeople Should Know and the Public Should Expect*, 3rd ed., Three Rivers Press, 2014, 9.
7   Ibid., 29.
8   Ibid., chapter 1.
9   Ross Gelbspan, *Boiling Point: How Politicians, Big Oil & Coal, Journalists, & Activists Have Fueled a Climate Crisis – & What We Can Do to Avert Disaster*, Basic Books, 2004, 79-80.
10 Ibid., 223.
11 Stephen Lacey, "American Newspapers Are Number One in Climate Denial," *Climate Progress*, 14 October 2012.
12 Dana Nuccitelli, "Conservative media bias is inflating American climate denial and polarization," *The Guardian*, 6 September 2016.
13 Denise Robbins, "It's Not Just The Editorial Page: Study Finds WSJ's Reporting On Climate Change Also Skewed," *Media Matters*, 11 August 2015.
14 Kevin Kalhoefer et al., "ANALYSIS: Wall Street Journal Opinion Section Is Chief Apologist For Exxon's Climate Change Deceit," *Media Matters*, 1 September 2016.
15 Denise Robbins, "STUDY: Newspaper Opinion Pages Feature Science Denial And Other Climate Change Misinformation," *Media Matters*, 1 September 2016.
16 Joanna B. Foster, "Poll: U.S. Leads the World . . . in Climate Denial," *Climate Progress*, 22 July 2014.
17 Pew Research Center, "What the world thinks about climate change in 7 charts," 18 April 2016.
18 Bill Kovach and Tom Rosenstiel, "The Elements of Journalism," 28.
19 Robert A. Hackett, et al., *Journalism and Climate Crisis: Public Engagement, Media Alternatives*, Routledge, 2017, Introduction.
20 Slezak, Michael, "The Great Barrier Reef: a catastrophe laid bare," *The Guardian*, 7 June 2016.
21 Hackett et al., *Journalism and Climate Crisis*.
22 Edward Maibach et al., *A 2016 National Survey of American Meteorological Society Member Views About Climate Change: Initial Findings*, George Mason University, Center for Climate Change Communication, March 2016.
23 J. Cook et al., "Consensus on consensus: a synthesis of consensus estimates

on human-caused global warming," *Environmental Research Letters* Vol. 11, No. 4, 13 April 2016.

24   David Edwards, "Analysis: Only one meteorologist in entire US linked 'climate change' to record hot Christmas," *Raw Story*, 29 December 2015.

25   C-Change Conversations, "Game Changers In Climate Change. A conversation with Steve Pacala," 27 February 2017.

26   Robert Cox and Phaedra P. Pezzullo, *Environmental Communication and the Public Sphere, 4th ed.,"*

27   Ellie Shechet, "Possible Trump Science Advisor Compares Climate Science to ISIS, Tells Us Jezebel 'Is Well Named,'" *The Slot*, 28 March 2017.

28   Kevin Kalhoefer, "Networks Covering March For Science Provided Platform For Climate Deniers," *Media Matters*, 24 April 2017.

29   "Bill Nye Destroys climate change-denying Trump adviser William Happer," 22 April 2017 (http://climatestate.com/2017/04/24/bill-nye-destroys-climate-change-denying-trump-adviser-william-happer).

30   West's Encyclopedia of American Law, "Negligence," The Gale Group, 2005.

31   Ibid., "Criminal Negligence."

32   Tom Engelhardt, "The Biggest Criminal Enterprise in History," TomDispatch, 23 May 2013.

33   Ibid.

34   Kevin Kalhoefer, "How Broadcast Media Covered Climate Change in 2016," *Media Matters*, 23 March 2017.

35   Tom Engelhardt, "Ending the World the Human Way: Climate Change as the Anti-News," 2 February 2014.

Chapter Five

CORPORATE AND BANK CRIME

Introduction

The concept of crime rests squarely on the shoulders of justice: on what is just and right, and what is unjust and wrong. This elemental sense within the human psyche has guided the law-making of civilization for millennia.

In July 2017, when the dangerous, climate-change-denying Trump Administration took the US out of the UN Paris agreement, a professor of law explained why climate disruption is a more profound moral issue than any other environmental problem. Central to this discussion is the fact that greenhouse gases from industrial nations do not disappear into thin air but are "long-lived" gases in the atmosphere, as recorded in the First IPCC assessment in 1990.

As has been known by governments for many years, climate impacts will most severely harm developing nations with catastrophic threats to their peoples' lives and the ecological systems that sustain them.

Professor Donald A. Brown wrote:

> It is the Trump assertion that the United States can base its energy policy primarily on putting US economic interests first while ignoring US obligations to not harm others that most clearly

provokes moral outrage around the world. The moral principle that people may not harm others on the basis of self-interest is recognized by the vast majority of the world's religions and in international law under the "no harm principle". The "no-harm" rule is a principle of customary international law whereby a nation is duty-bound to prevent, reduce, and control the risk of environmental harm to other nations caused by activities within the nation.[1]

A second widely accepted principle derives from the U.N. *Declaration on the Rights of Indigenous Peoples*, requiring companies and governments to obtain free, prior, and informed consent from Indigenous Peoples before doing business on their land. Article 29.1 of the Declaration states that "Indigenous peoples have the right to the conservation and protection of the environment and the productive capacity of their lands or territories and resources."

A third principle is embodied in the 1948 *UN Declaration of Human Rights*, which set out fundamental human rights to be universally protected. This *Declaration* has been translated into over 500 languages. It states that:

- Everyone has the right to life, liberty and security of person (Article 3).
- Everyone has the right to a standard of living adequate for the health and well-being of himself and of his family (Article 25).

A fourth set of principles appeared on June 16, 2011, when the United Nations Human Rights Council unanimously endorsed the *Guiding Principles for Business and Human Rights*, a global standard for preventing and addressing the risk of adverse impacts on human rights linked to business activity.

An article by law professor Lynda Collins discusses the progressive integration of international human rights law with international environmental law. She examines the violation of recognized human rights in terms of environmental deprivation:

> Taking the right to life, for example, it is not necessary to formulate a new "environmental component" of the right to life in order to address lethal environmental harm. Instead, courts need only recognize that environmental harm may cause loss of life just as surely as other means. If a citizen is asphyxiated by noxious gases emanating from a government-operated incinerator, she is equally dead as if she had been shot or beaten by government agents. It would be irrational for human rights law to provide less [sic, more] protection in the latter scenario than it does in the former; this would, in a sense create an environmental exemption from the right to life.[2]

In December 2015, at the UN Paris Climate Conference, based on the evidence of the IPCC, 195 nations agreed to aim for a global warming limit of 1.5°C, recognizing that this would significantly reduce the risks and impact of climate change. As we have seen, this agreement means that continued financing and production of fossil fuel energy must stop.

Within the context of the United Nations' enshrined principles above, we consider the continued financing and production of fossil fuel energy, and the egregious denial of climate change by politicians, as gross violations of basic human rights and therefore as the most serious possible crime against humanity. In particular, it is a crime against children alive today, and against the future unborn.

Bank Crime

Every year since 2009, the Sierra Club, Rainforest Action Network, BankTrack, and Oil Change International have produced a joint research analysis called "Shorting the Climate: Fossil Fuel Finance Report Card."

In 2017, the "Report Card" rated the financing by 37 major private banks in each of the most carbon-intensive "extreme" fossil fuels sectors: coal mining, coal power, extreme oil (tar sands, Arctic oil, ultra-deep drilling) and North American liquefied natural gas export. The ratings included banks in the US, Canada, Europe, Japan, China and Australia, which together funneled $87 billion to fund the production of extreme fossil fuels in 2016, down 22% from the previous year.[3]

From 2014 to 2016, the major banks poured $105 billion into Arctic oil, the tar sands, and ultra-deep offshore oil, led by the Royal Bank of Canada and JPMorgan Chase. Chase also led LNG exports in North America, to the tune of $51 billion

In Maryland and Pennsylvania, massive multi-bank investments in LNG exports are driving huge export facilities in densely populated areas, and contaminating drinking water and the quality of life generally.

Even at the World Bank $5 billion in development policy loans have been creating subsidies for coal, gas and oil projects, and undercutting initiatives to build wind, solar and geothermal power infrastructure and protect vulnerable rainforests, including in the Amazon[4]

How big banks are "shorting the climate" is explained below:

> In finance industry terms, "short-selling," or shorting, is a transaction through which an investor profits if a company or asset declines in value. After Paris, financing fossil fuels is tantamount to shorting the climate. Financial institutions that support business-as-usual for the fossil fuel industry are placing their bets on companies whose long-term success depends on runaway climate change.[5]

To skirt this ultimate morality issue, the Thun group of banks (including Barclays, Credit Suisse, Deutsche Bank, and JPMorgan Chase) proposed in a discussion paper that banks should be judged to be meeting the standards of the "U.N. Guiding Principles on Business and Human Rights," claiming that banks are not responsible for the human rights impacts caused by their clients' dirty energy projects (such as displacing communities). The banks, by failing to see that they were directly enabling thousands of such projects, clearly sought to deny the purpose of the UN principles. The discussion paper of the banks was met with widespread public derision.

In 2016, the most glaring North American example of human rights violations enabled by banks was the financing of the Dakota Access Pipeline (DAPL) and the gross violations of indigenous rights it caused.

Coal, meanwhile, is widely recognized as the most toxic and emissions-intensive fossil fuel. The leading worldwide cause of climate change, coal-fired power plants are the largest source of CO_2 – greater than tropical deforestation or oil use for transportation. An immediate rapid transition away from coal-fired energy is essential to limiting global warming to 1.5°C (or the catastrophic 2°C).

While banks continue to highlight their investments in clean energy, at least $500 billion in bank financing has funded hundreds of coal companies since 2005.[6] Between 2014 and 2016, 37 of the biggest international banks provided $57.92 billion for coal mining and $74.71 billion for coal power, with the Chinese megabanks at the top of the list.[7]

The Canadian, Japanese, Chinese, and Australian banks fared worst on coal power in the 2017 "Shorting the Climate" rankings. The US banks were somewhat better, and the European banks were best (but all remain unconscionable).

The top eight coal power financiers were Citigroup, Barclays, JPMorgan Chase, Bank of America, Royal Bank, Wells Fargo, and Morgan Stanley. In 2016, JPMorgan Chase, while paying lip service to the Paris climate accord, poured $6.9 billion into the dirtiest fossil fuels on earth, and was top banker on Wall Street in tar sands oil, Arctic oil, ultra-deep water oil, coal power, and LNG export.[8]

President Trump has pledged to revive the coal industry and to roll back its regulations. This will encourage more mountaintop removal, a method of strip coal mining that is using explosives to blast up to 1000 feet off the beautiful mountains of America's Appalachian Shield, creating havoc and misery for dozens of communities. This too has been enabled by banks.

Nine coal mega-mines have been proposed for Australia's massive untapped Galilee Basin. At full production, they would rank as the world's seventh biggest contributor to CO_2 pollution, and would drive the expansion of coal ports along the Great Barrier Reef coast.

The global coalition of 28 NGOs that produces the "Fossil Fuel Finance Report Card" has been opposing these coal projects plus many extreme oil projects for years, causing the withdrawal of certain banks listed in their reports, and commitments to back out from other major banks. While this is seen as encouraging progress, the billions keep coming.

The mining and burning of coal is starkly incompatible with a stable climate, as are the extreme oil projects in the Arctic, tar sands, and deep in the oceans. There is no excuse for banks to support an industry that flies in the face of the IPCC science and the Paris commitments to limit global warming to 1.5°C.

The banks can choose which ventures they wish to enable, and that choice is an ethical one. They could move quickly and entirely to renewable financing. But preferential enabling of the grotesque, gaping mountaintop coal projects, the vast gaping tar sands, and the filthy contamination of hydraulic fracturing cannot be described as anything but criminal savaging of the planet, as they drive CO_2 sky high to afflict the futures of children being born today.

It is important that people look into where their money is going, and then switch their investments out of fossil fuels into clean renewable energy and switch their bank accounts to the most ethical climate-friendly banks and credit unions. As an example, U.S. Bancorp announced in April 2017 that it had pulled all its investments in pipelines and switched to wind and solar.

Corporate Crime

The global swathe of environmental and human rights disasters left by the oil industry is legion and the lawsuits have been legendary. Chevron in Ecuador, the Exxon Valdez, Gulf of Mexico.

Every October, petroleum executives hold an international conference titled "Oil and Money." In 2016, hosted by Energy Intelligence and *The New York Times,* it met in London with the theme "Boom, Bust and Beyond: Strategies for Survival."

With 2016 the hottest year on record, the question arises, "Whose survival?" The survival of oil execs with their multi-million dollar bonuses, or the survival of people worldwide traumatized by extreme weather events – from the heaviest snowstorm in New York since 1869, to killer floods in Texas, to deadly typhoons in the Philippines and Japan, to 115°F heat in Spain, to temperatures of 129°F in Iraq and Kuwait, to a drought in India causing $5 billion in damages.

With many municipalities, states, and NGOs struggling to meet the 1.5°C targets set in Paris, perhaps it was the brashness of the "Oil and Money" theme that triggered an organized protest in Hyde Park called "Oil & Money: This is a crime scene."

The naked immorality of corporate blindness to human suffering had galled both the protestors and the 60 prominent people who signed an open letter calling for the immediate closure of the conference.

This repugnance rested on the instinctive moral sense that not just the actions but the attitude of the oil industry is criminal. The awareness of the criminality of vast energy profits versus the survival of nature and humanity is receiving international legal recognition. A recent policy paper from the Office of the Prosecutor of the International Criminal Court addressed it in a paper dealing with case selection and prioritization:

Although the International Criminal Court (ICC) cannot directly hear climate crimes, the Office of the Prosecutor of the ICC has stated – in its recent policy paper on case selection and prioritization – that war crimes, crimes against humanity or genocide committed through, or resulting in, the destruction of the environment, the illegal exploitation of natural resources or the illegal dispossession of land, warrant particular attention when selecting cases for investigation and prosecution before the ICC. The Office of the Prosecutor will also support national authorities who investigate and prosecute these crimes.[9]

The United Nations, in discussing the elements of crimes against humanity, refers to the Rome Treaty, which states that these crimes "do not need to be linked to an armed conflict and can also occur in peacetime." They add:

In contrast with genocide, crimes against humanity do not need to target a specific group. Instead, the victim of the attack can be any civilian population, regardless of its affiliation or identity. Another important distinction is that in the case of crimes against humanity, it is not necessary to prove that there is an overall specific intent.[10]

Where is the Punishment?

The orchestrated crimes of state-corporate denial and disinformation going back to the 1980s – when industry and government knew that climate change was on its way – were crimes of commission, and were bad enough. (See Chapter 2).

But the actual devastation the big majors have visited upon Earth's landscapes – the oceans, rainforests, tundra, and most importantly, the atmosphere – as they plunder the world for fossil fuels is beyond telling. According to the 2010 "Carbon Majors" study, about 63% of cumulative GHG emissions since 1751 have come from only 90 investor and state-owned institutions.

The fines they pay for these toxic abuses are simply factored into the costs of doing business, and lawsuits drag on for years as corporations take every possible step to avoid facing the music.

Internal Chevron tapes show cynical employees searching for uncontaminated core samples to prove to a judge that the company had not polluted vast tracks of the Ecuadorian rainforest with toxic oil waste. A decade later, Ecuador is still seeking compensation from Chevron for its Supreme Court judgment of $9.5 billion.

The dumping of 18 billion gallons of oil waste into Ecuador's ground-water and rivers has spurred an unprecedented public health epidemic of miscarriages, birth defects and childhood leukemia – and also the loss of farm animals, agriculture and fishing.

The suffering of Ecuador's children is nothing now compared to what it will be for the world at 2°C, or much higher, if emergency action is not launched immediately. The frenzy for cheaper oil has been a corporate crime against humanity since companies like Exxon knew in 1980 that climate change is real,[11] and it's been happening all over the world. Exxon is being sued by 18 state Attorneys-General for violating consumer protection laws with misleading fraudulent statements about climate change.[12]

Air pollution, which is strongly linked to carbon-based energy production, kills about 9 million people a year, according to a 2017 report by The Lancet Commission on Pollution and Health.[13] China's emissions in particular are crucial to the success of global efforts to reduce emissions. Its problems are our problems.

In Africa particularly, extraction, land clearing and pipelines undertaken by major oil companies – including Shell in Nigeria, Canada's Talisman (now defunct) and Sweden's Lundin

Petroleum in Sudan – are claimed by human rights monitoring agencies to have fueled conflict and civil war. The rationalization has been, "Sure, we do it, but so does everybody else."

And they have known all along, since Jim Hansen testified before Congress, of the criminal impact that the future would hold. In 1989 Shell, for instance, announced it was redesigning an oil platform in the North Sea to account for rising sea levels and worsening storms connected to climate change.

They cannot claim they did not know that burning oil was causing these problems. The idea that climate change was caused by greenhouse gases had been understood at least since 1957, when physicist Edward Teller presciently told an assembly of scientists that rising CO2 levels might eventually melt back the polar ice caps and inundate the world's lowlands.[14]

In 1998, at a time when toxic abuses were clearly manifested abroad, the American Petroleum Institute wrote in an action plan:

> Unless "climate change" becomes a non-issue, meaning that the Kyoto proposal is defeated and there are no further initiatives to thwart the threat of climate change, there may be no moment when we can declare victory for our efforts.[15]

The question today is not whether but how long the highly politicized rogue fossil fuel industry will continue to seek victory against the truth of climate change and to ravage human rights in spite of emergency climate disruption that is becoming visible to all.

It's no accident that although the media is now paying continual lip service to the effects of climate change, none are speaking of its causes or of an "emergency," which requires a tipping point in both awareness and willingness to leave business-as-usual behind and to act with decisive purpose to manage a full transition to renewable energy.

Unfortunately, industry, government, and the media are not yet ready to give up their profits and lifestyles for the sake of their children. But once the emergency is declared and solid leadership has emerged, citizens will willingly follow.

Endnotes

1   Donald A. Brown, "The Moral Outrageousness of Trump's Decision on the Paris Agreement," *Ethics and Climate (blog)*, 7 July 2017.

2   Lynda Collins, "Are We There Yet? The Right to Environment in International and European Law," *McGill International Journal Sustainable Development Law & Policy*, Vol. 3, No. 2, 119-153, 2007.

3   Rainforest Action Network et al., *Fossil Fuel Finance Report Card 2017: Banking on Climate Change*, 21 June 2017. This report was in collaboration with 27 other NGOs.

4   Bank Information Center, *World Bank introducing new fossil fuel subsidies, undermining its own climate change commitments and forest protection efforts*, 26 January 2017.

5   Rainforest Action Network, *Shorting the Climate: Our New Report*, 2016.

6   BankTrack, *Banks: Quit Coal!* (http://coalbanks.org/) and *The Top Twenty Coal Banks* (http://coalbanks.org/#score).

7   BankTrack, *Banks and Coal*, 24 July 2017.

8   Sierra Club, *New Report Reveals Bank Policies Fail to Respond to Climate Risks,* 21 June 2017.

9   Center for Climate Crime Analysis, *Prosecuting Crimes as a Matter of Priority and General Deterrence* (http://www.climatecrimeanalysis.org/priority-prosecution.html). This Center is run by Reinhold Gallmetzer, Appeals Counsel at the Office of the Prosecutor of the International Criminal Court, Netherlands. The CCCA is a non-profit organization of prosecutors and law enforcement professionals to trigger and support prosecutions of climate crimes that result in significant amounts of greenhouse gases. It has been endorsed by Interpol. See also the ICC original document at https://www.icc-cpi.int/itemsDocuments/20160915_OTP-Policy_Case-Selection_Eng.pdf.

10   United Nations. Office on Genocide Prevention and the Responsibility to Protect, *Crimes Against Humanity: Definition,* n.d., cited September 7, 2017 (http://www.un.org/en/genocideprevention/crimes-against-humanity.html).

11   Neela Bannerjee et al., "Exxon: The Road Not Taken," *Inside Climate News*, 15 September 2015.

12   See https://www.cnsnews.com/news/article/barbara-hollingsworth/11-state-attorneys-general-side-exxonmobil-climate-change-case.

13   Philip J. Landrigan et al., "The Lancet Commission on Pollution and Health," 19 October 2017 (http://www.thelancet.com/commissions/pollution-and-health).

14   American Institute of Physics, *The Discovery of Global Warming*, January 2017.

15   American Petroleum Institute, *Global Climate Science Communications Action Plan*, 1998 (https://insideclimatenews.org/documents/global-climate-science-communications-plan-1998)

Chapter Six

MORAL COLLAPSE AND RELIGIOUS APATHY

In late 2014, a year before the Paris Conference, Cambridge University Press published ethicist Clive Hamilton's essay, "Moral Collapse in a Warming World."

In it he stated some straightforward truths that are seldom heard in politics, the media, or even around the dinner table:

> Carbon emissions have now accumulated to the point where avoiding warming by 2 degrees Celsius is impossible. Even under optimistic assumptions about the speed with which countries might respond, it now seems likely that the world will warm by 4 degrees Celsius or more, which will transform the conditions of life on the planet and result in catastrophes. Even the overly cautious analysis of the Intergovernmental Panel on Climate Change (IPCC) cannot disguise these facts. Indeed, we know with certainty what must be done to avoid enormous harm, particularly to poor and vulnerable people—a task rendered less onerous by the fact that all economic studies show that the transition to a low-carbon energy system could be achieved at modest cost. Rather than creating a perfect storm, the ethical

winds blow strongly in one direction. Yet moral corruption prevails not because the situation is inherently murky, but because confusion has been deliberately sown, and because the public and political representatives have welcomed reasons to shirk their ethical obligations.

There are three kinds of actors in this process of subversion: those who tell the lies, those who repeat the lies, and those who allow themselves to be seduced by the lies.[1]

Moral Disengagement and Collapse

We are dealing with the phenomenon that even though climate change is a severe threat to future civilization, this circle of lies and subversion causes inaction at all levels.

To explore this incongruity, we turn to the ideas of moral agency and moral disengagement:

> A moral agent is a person who has the ability to discern right from wrong and to be held accountable for his or her own actions. Moral agents have a moral responsibility not to cause unjustified harm.[2]

A "moral agent" therefore is someone who is capable of choosing to do things rightly or wrongly. Moral agency is reflected in the power to behave humanely or inhumanely.

The term "moral disengagement" has been applied to the baffling inaction to cope with the growing climate crisis. Moral disengagement is

> a term from social psychology for the process of convincing the self that ethical standards do not apply to oneself in a particular context.

This is done by separating moral reactions from inhumane conduct and disabling the mechanism of self-condemnation.[3]

To put it plainly, when people disable their moral reactions to inhumane conduct in any context, they enter a state of moral collapse.

Moral Collapse: Industry, Government, and Media

Since the 1980s, the international science community, based on its overwhelming evidence, has been struggling to meet the challenge of climate change.

How Big Oil has systematically defeated these efforts is revealed in a concise video by Dr. Donald A. Brown, Professor of Ethics and Law at Widener University. Brown unwraps an ethically abhorrent, reckless disregard for truth that constitutes a new kind of crime against humanity requiring a new kind of legal response.[4]

A second short video, which won the CinemAmbiente 2017 award in Turin, documents Exxon scientists' written statements advising their board, from 1977 to 1989, of the atmospheric dangers caused by burning fossil fuels. In 1996 ExxonMobil began its 20-year disinformation campaign to publicize the opposite story.[5]

Philosophy Professor Dr. Lawrence Torcello has described the criminal negligence of funding climate denial:

> We have good reason to consider the funding of climate denial to be criminally and morally negligent. The charge of criminal and moral negligence ought to extend to all activities of the climate deniers who receive funding as part of a sustained campaign to undermine the public's understanding of scientific consensus.[6]

The fossil-fuel companies and the governments that help them with subsidies and tax shelters had by 2016 ignored the *average annual displacement of 21.5 million people since 2008* from climate and extreme weather events.[7] Under the Rome Statute of the International Criminal Court, "deportation or forcible transfer of population" is considered to be a crime against humanity.

By failing to protect people from industry lies, the corporate media and the governments of the United States and other countries have been the most obvious accomplices in enabling immoral and inhumane policies for economic gain.

Moral Disengagement

The climate change crisis finally began sweeping the North American news in 2015, the year of the Paris summit, yet many people continue to "allow themselves to be seduced by the lies," and to justify business-as-usual consumption. This is shown by the growing market share of SUVs and pickup trucks:

> The US auto market continued its hot streak in 2016 with more than 17.4 million cars and trucks sold. As expected, pickup trucks and SUVs accounted for much of the growth ... Compact-car sales fell 6.1% last year. At the same time, its mid-size counterpart saw sales plummet 11.2%.[8]

As the authors of "Climate Change and Individual Responsibility" wrote in 2015:

> Despite overwhelming evidence to the contrary, the harmful effects of climate change are outrightly *denied* or blamed on natural processes, scientific uncertainties are overly emphasized and it is alleged that *it won't be that bad.* Many claim either ignorance or that *it is not* their *fault.*

Regarding their engagement in greenhouse gas emitting activities, emitters maintain that *it makes no difference whether they do or don't* and that *any way, everybody does it.*"[9]

Probing these attitudes, the authors describe a tendency "to employ psychological mechanisms of *moral disengagement* by which people reconstruct their moral judgment of climate change and their contribution to it in order to evade individual responsibility."[10]

Moral Disengagement in Global Negotiations

In June 2017, responding to the Trump administration's decision to pull out of the UN Paris Agreement, Archbishop Marcelo Sanchez Sorondo, Chancellor of the Pontifical Academy of Sciences, called it "a great evil and irrational," because it goes against science and the indisputable evidence that climate change "is impacting on the health of people across the globe."[11]

Philosophy professor Stephen Gardiner depicts climate change as "a perfect moral storm," citing a form of "moral schizophrenia" in the face of "inconvenient duties."[12]

The complexity of the problem, he suggests, may be perfectly convenient for our current generation, and for each successor generation when it comes to negotiating international climate accords:

> For one thing it provides the cover under which it can seem to be taking the issue seriously – by negotiating weak and largely substanceless global accords, for example, and then heralding them as great achievements – when really it is simply exploiting its temporal position.[13]

In the media, climate change negotiations do not hold a candle to the coverage of U.S. deaths from terrorism (one terrorism death per 1,000 deaths from firearms[14]), which are trumpeted with consternation daily, giving enormous satisfaction and encouragement to the perpetrators.

Yet it remains indisputable that global warming is the primary moral harm violating the basic existential rights of all sentient beings.

Basic Human Rights

In referring to the Universal Declaration of Human Rights, which was approved by the UN General Assembly in 1948, Prof. James Nickel wrote that it referred to rights that are "universal, to be held by people simply as people."

In 1789 the French Declaration of the Rights of Man spoke of rights that are "natural" and hence *"imprescriptible."* The American Declaration of Independence spoke of "unalienable rights," and philosopher Alfred North Whitehead spoke of "the essential rights of human beings, arising from their sheer humanity."

The 1980 Presidential Commission on World Hunger spoke of basic rights: "Whether one speaks of human rights or basic human needs, the right to food is the most basic of all. Unless that right is first fulfilled, the protection of other human rights becomes a mockery."[15]

The right to fresh water is equally necessary for life, as is the right to non-toxic air. Basic rights, by virtue of being basic, trump all non-basic rights.

For those who claim the right to large profits and luxurious lifestyles, insofar as it exists – it is really a desire, not a right – it would be immoral for opulence to trump the vital necessities, and hence the basic rights, of others. John Locke, whose writings inspired much in early American thought, "had taken for granted that the right to accumulate private property was limited by a universal right to subsistence."[16]

In light of these principles, the "right" of ExxonMobil, Chevron, and the Koch brothers to accumulate vast sums of money would, in a moral system of national and global governance, be trumped by the basic rights of people around the world to clean air and water for drinking and agriculture, to thriving land and marine life, and to a stable sea-level allowing them to remain in their homes.

Intergenerational Justice

The first principle of intergenerational justice is that the rulers of nation-states are understood to morally represent the interests of their citizens in perpetuity.

The most famous statement about intergenerational justice is the "Great Binding Law" in the *Constitution of the Iroquois Nations*, which says:

> In all your official acts, self-interest shall be cast into oblivion. . . . Look and listen for the welfare of the whole people and have always in view not only the present but also the coming generations, even . . . the unborn of the future Nation.[17]

Intergenerational justice is the moral principle most cited by climate scientists.

- "The basic matter," James Hansen has said, "is not one of economics. It is a matter of morality – a matter of intergenerational justice. The blame, if we fail to stand up and demand a change of course, will fall on us, the current generation of adults."[18]
- "As a father of a six-year-old daughter," Michael Mann wrote in 2012, "I believe we have an ethical responsibility to make sure that she doesn't look back and ask why we left her generation a fundamentally degraded planet relative to the one we started with."[19]

Our current inaction on climate change is intergenerational theft. We are borrowing from the prosperity of our children – sacrificing their prosperity and safety for a mere one to two percent of current world GDP. In doing so we are creating problems for them that are far easier to fix now than in 20 or 30 years' time.

Responsibility for CO2 Emissions

A principle of climate ethics is that the countries most responsible for global warming, who have benefited most from the burning of fossil fuels, need to accept primary responsibility to pay for mitigation and adaptation.

As we have seen, global warming has already unjustly cursed the lives of millions of displaced people in southern countries:

> The irony with climate refugees rests in the fact that those being forced to leave their homes and land have played close to no role in increasing the rate of climate change. The level of greenhouse gases emitted by these developing countries is close to nothing in comparison to developed countries.[20]

At the 2009 climate conference in Copenhagen, the developed countries promised to mobilize $100 billion a year of climate financing for developing countries by 2020.

The aim was to help poorer countries adapt to change, and as they developed, to bypass fossil fuels in favor of cleaner alternatives. Much of the money was to flow through the Green Climate Fund, a U.N. body that was later set up to administer the funds.

The Paris climate accord of 2015 reaffirmed this commitment, extended it to 2025, and urged developed countries to come up with a "concrete roadmap."

But as of 2017, developed countries had committed only $60 billion, much of which had been previously allocated aid and was not new funding.[21]

Equity in Fiscal Responsibility

A November 2014 report from the World Resources Institute shows that China has the greatest CO2 emissions today. However, we have seen that climate change is mainly driven by locked-in cumulative emissions over time.

From 1850 to 2011, the United States had contributed 27% of total world emissions; Europe's were 25%, and combined they were 52%. This compared to 11% for China, 8% for Russia, 3% for India, and 26% for the rest of the world.[22]

Equity of payment to prevent worse global warming in future can be discussed in terms of responsibility for cumulative emissions over time. But thus far Western nations, especially the USA, have failed to translate their emissions debt into concrete financial action.

Divestment Responsibility

Divestment is the selling off of fossil fuel stocks, bonds, or other holdings, because, in Bill McKibben's words, "if it's wrong to wreck the climate, it's wrong to profit from that wreckage."

The power of divestment lies in the ability of the investor (whether a university, pension fund, municipality, foundation or individual) to shrink the fossil fuel economy.

Consensus is growing that owning oil securities is as ethically repugnant as profiting from tobacco stocks. By December 2016, the value of investment funds committed to selling off fossil fuel assets had jumped to $5.2 trillion, having doubled in just over a year.

A 2017 book, *The Clean Money Revolution*, estimated that by 2050, $40 trillion will change hands from Boomers to Millennials – the largest intergenerational wealth transfer in history. The "dirty money" of business-as usual now hangs in the balance as society heads for an investment revolution.[23]

Fulbright scholar Alex Lenferna reflects on the ethical power investors have, even at this late date:

By divesting, institutions and individuals can play
an important leadership role by acting on the deep
moral urgency of the climate crisis, and investing
instead in a prosperous low-carbon future, which
prevents the worst ravages of climate change, at
least while it is still possible to do so.[24]

Abolition

In 1833, after the industrial revolution, the slave trade was
abolished in England, and the U.S. followed suit in 1865. At
about the same time, oil was emerging to replace slavery as the
underpriced source of abundant energy.

There are striking parallels between the early struggles
against slavery and the current struggle to abolish fossil fuels:

- "Just as few people saw a moral problem with
  slavery in the 18th century, few people in the 21st
  century see a moral problem with the burning of
  fossil fuels."[25]
- Insofar as they *are* aware of the moral problem,
  "most in our society simply cannot conceive of a
  way for our economy to be powered by anything
  other than fossil fuels."[26] Similarly, those who
  depended on slaves could not see any other
  solution to their energy needs.
- "US Congressmen tend to rationalize fossil fuel
  use despite climate risks to future generations
  just as southern congressmen rationalized slavery
  despite ideals of equality."[27]
- The ease and wealth provided by slaves once gave
  people an incentive not to act on awareness of the
  immorality of slavery; today's "energy-slaves"
  give contemporary society strong incentive not to
  act on climate change.
- "In the same way that the abolition movement

cast a shadow over the cotton boom, so does the movement to put a price on carbon spook the fossil fuel companies."[28]

- "Because the abolitionists were ultimately successful, it's all too easy to lose sight of just how radical their demand was at the time: that some of the wealthiest people in the country would have to give up their wealth. That liquidation of private wealth is the only precedent for what today's climate justice movement is rightly demanding: that trillions of dollars of fossil fuel stay in the ground."[29]

Just as the moral arguments for abolishing slavery remained ineffectual while there seemed little alternative, the moral and scientific arguments to abolish fossil-fuel energy for years fell on deaf ears.

However, thanks to entrepreneurial innovation, the boundless supply of wind and sunlight has gradually become more competitive than coal, oil, and natural gas, as will be seen in Chapters 9 and 10. Social awareness of this emerging reality has been generally suppressed by the (morally-disengaged) corporate networks bowing to powerful petroleum interests.

Religious Apathy

On June 18, 2015, Pope Francis published his now-famous *Encyclical Letter* on "care for our common home."

It began with the words, "Now, faced as we are with global environmental deterioration, I wish to address every person living on this planet."[30]

Central to his message is the world's need for an "an integrated approach to combating poverty, restoring dignity to the excluded, and at the same time protecting nature." A true ecological approach, he said, "must integrate questions of justice in debates on the environment, so as to hear both the cry of the earth and the cry of the poor."

These cries originate from one transgression:

> The violence present in our hearts, wounded by sin, is also reflected in the symptoms of sickness evident in the soil, in the water, in the air and in all forms of life. This is why the earth herself, burdened and laid waste, is among the most abandoned and maltreated of our poor; she "groans in travail" (*Rom* 8:22). [31]

In the days that followed, Protestant, Jewish, Buddhist and Muslim leaders of the world stood with Pope Francis. The Archbishop of Canterbury issued the Lambeth Declaration, calling on faith communities worldwide to transition to a low carbon economy.

The Dalai Lama tweeted that "since climate change and the global economy now affect us all, we have to develop a sense of the oneness of humanity."

Imam Mohamed Magid of the Islamic Society of North America said that "People of all faiths can come together for this cause because the concept of stewardship on Earth is a shared belief."[32]

Rev. Sally Bingham, founder of Interfaith Power & Light, said it best:

> The impact of today's message is clear: if you are a person of faith, you have a responsibility to address climate change. It's as simple as that.[33]

The importance of this consensus was noted by Rabbi Fred Scherlinder Dobb, chair of the Coalition for the Environment and Jewish Life:

> Many government and industry leaders note the religious unity around this issue. If the world's religions can agree, surely the nations of the

world can do the same, and then the real work on climate change can begin.[34]

Indeed, about 80 percent of the world's people identify with a religious faith. Why then, three years later, have world faith leaders and their followers not powerfully moved society towards a near zero carbon future?

Theologian Paul Tillich, who left Germany in protest when Hitler came to power in 1933, famously defined religion as "ultimate concern." Given that definition, religion deals with matters of ultimate concern, and matters of ultimate concern are by definition religious. Being religious is being grasped by an ultimate concern.[35]

The opposite of concern is apathy, which still blocks resolute emergency action in most communities, including the faith community.

A Moral Imperative for Climate Justice

Pope Francis, perhaps aware of this apathy, advanced a moral imperative in his 2015 *Encyclical Letter*:

> We are always capable of going out of ourselves towards the other. Unless we do this, other creatures will not be recognized for their true worth; we are unconcerned about caring for things for the sake of others; we fail to set limits on ourselves in order to avoid the suffering of others or the deterioration of our surroundings. Disinterested concern for others, and the rejection of every form of self-centeredness and self-absorption, are essential if we truly wish to care for our brothers and sisters and for the natural environment. These attitudes also attune us to the moral imperative of assessing the impact of our every action and personal decision on the world around us.[36]

In September 2016, following the hottest year on record, Francis issued a yet more urgent cry to protect "those least responsible for climate change," calling climate degradation "a sin against God."[37]

The only mainstream U.S. newspaper to report the Pope's desperate call for humanity was the *Washington Times.*

Once again the corporate media, blind to human suffering beyond U.S. borders, chose servitude to the fossil fuel industry, the Koch brothers, and the Republican Party.

This conscious morally-destitute choice leaves no room for continued apathy by people of faith as a response to global warming. Religious leaders and their communities must embrace "the ultimate concern" and protect the climate as the most indispensable common good.

This means calling for a transition away from fossil-fuel based business-as-usual. This in turn means calling for the immediate long-promised termination of pernicious government fossil fuel subsidies, which amount to trillions of U.S. dollars worldwide each year.

Endnotes

1   Clive Hamilton, "Moral Collapse in a Warming World," EIA: Ethics & International Affairs, 4 September 2014.

2   "Ethics Unwrapped" (http://ethicsunwrapped.utexas.edu/glossary/moral-agent).

3   Susan T. Fiske, "Social Beings: Core Motives in Social Psychology," Wiley, 2004.

4   Donald A. Brown, "Why The Climate Change Disinformation Campaign Is So Ethically Abhorrent," ethicsandclimate.org, 9 August 2012 (http://blogs.law.widener.edu/climate/2012/08/09/why-the-climate-change-disinformation-campaign-is-so-ethically-abhorrent).

5   Hogan Millar Media, "Oil Slick," 2016 (https://vimeo.com/172761267).

6   Lawrence Torcello, "Is misinformation about the climate criminally negligent?"*The Conversation,* 13 March 2014.

7   UN Refugee Agency, "Frequently asked questions on climate change and disaster displacement," 6 November 2016.

8   Benjamin Zhang, "The 20 Best-Selling Cars and Trucks in America," *Business Insider*, 7 January 2017.

9   Wouter Peeters et al., "Climate Change and Individual Responsibility:

Agency, Moral Disengagement and the Motivational Gap," Palgrave MacMillan, from the Introduction, 2015.

10   Ibid.

11   Gerard O'Connell, "Vatican Official: President Trump's Decision to Withdraw from Paris Climate Change Accord a 'Disaster,'" *American Magazine*, 2 June 2017.

12   Stephen M. Gardiner, "A Perfect Moral Storm: The Ethical Tragedy of Climate Change," Oxford University Press, 2011, 339.

13   Ibid., 408-09.

14   Eve Bower, "American deaths in terrorism vs. gun violence in one graph," *CNN*, 3 October 2016.

15   Henry Shue, *Basic Rights*, quoting John Stuart Mill, *Utilitarianism* Indianapolis: Bobbs-Merrill, 1957, 67; Philip Alston, "International Law and the Right to Food," in Asbjorn Eide et al., ed., *Food as a Human Right*, The United Nations University, 1984, 162-74, at 162.

16   Shue, *Basic Rights*, 153, referring to John Locke, *Two Treatises of Government,* Book II, Ch. V.

17   The Constitution of the Iroquois Nations: The Great Binding Law.

18   James Hansen et al., "The Case for Young People and Nature: A Path to a Healthy, Natural, Prosperous Future," From James Hansen (Blog), 5 May 2011.

19   Michael Mann, "The Danger of Climate Change Denial," Climate Progress, 23 April 2012.

20   Anam Sultan, "Mass Migration: The Untold Crisis of Climate Refugees," *Earth Reform,* 4 August 2012.

21   Lucy Hornby, "Signatories to climate change agreement in dispute over financing," *Financial Times,* 11 April 2017.

22   Mengpin Ge, Johannes Friedrich and Thomas Damassa, "6 Graphs Explain the World's Top 10 Emitters," World Resources Institute, 25 November 2014.

23   Joel Solomon and Tyee Bridge, "The Clean Money Revolution: Reinventing Power, Purpose, and Capitalism," New Society Publishers, 2017 (https://www.publishersweekly.com/978-0-86571-839-5).

24   Alex Lenferna, "Divest-Invest: A Moral Case for Fossil Fuel Divestment," 2016, draft chapter from an upcoming collection, *Climate Justice: Economics and Philosophy*, to be published by Oxford University Press, Henry Shue and Ravi Kanbur, eds.

25   Andrew J. Hoffman, "By Invitation: Climate Change: Calling the Fossil Fuel Abolitionists," Ethical Corporation, 28 March 2008.

26   Leah D. Schade, "Fossil Fuel Abolitionists," *Ecopreacher*, 9 November 2012.

27   Ibid.

28   Chris Hayes, "The New Abolitionism," *The Nation*, 22 April 2014.

29   Ibid.

30   "Encyclical Letter, Laudato Si' of the Holy Father Francis, On Care For Our Common Home," No. 54, May 2015.

31   Ibid., para 2.

32   Aisha Bhoori, "Faith Leaders Stand With Pope Francis on Climate Change," *Time,* 17 June 2015.

33   Interfaith Power & Light, "Faith Leaders Praise Pope Francis' Historic Encyclical, Defend Against Backlash," 18 June 2015 (http://www.interfaithpowerandlight.org/2015/06/faith-leaders-praise-pope).

34   Ibid.

35   D. Mackenzie Brown, ed., *Ultimate Concern: Tillich in Dialogue*, Harper &
        Row, 1965.
36   "Encyclical Letter, Laudato Si' of the Holy Father Francis, On Care For Our
        Common Home."
37   Bradford Richardson, "Pope Francis Calls Climate Change a 'Sin,'"
        Washington Times, 1 September 2016.

PART II

GAME CHANGERS
FOR SURVIVAL

Chapter Seven

ENERGY SUBSIDIES AND TAX REFORM

We begin this first chapter in Part II (which presents solutions to the problems outlined in Part I) with fossil fuel subsidies – the most absurd and destructive energy policy and the easiest to correct.

Market manipulation in the form of fossil fuel subsidies is a tool governments use to persuade the market to prefer these fuels. It is now a matter of the survival of civilization that these energy subsidies be transferred to renewables.

Fossil Fuel Subsidies

A fossil fuel subsidy has been defined as "any government action that lowers the cost of fossil fuel energy production, raises the price received by energy producers, or lowers the price paid by energy consumers. Essentially, it's anything that rigs the game in favor of fossil fuels compared to other energy sources."[1] Total direct subsidies by all nations ran close to $500 billion in 2014[2] – but this astonishing figure represents only 10% of the full cost to society of using fossil fuels for energy.

In 2015, an International Monetary Fund (IMF) working paper estimated that the world's governments would spend a staggering $5.3 trillion on energy subsidies that year, with most of it supporting fossil fuels. This worked out to $10 million a minute, more than the total amount world governments would spend on health.[3]

The five IMF authors had taken a new approach to measuring the true cost of fossil fuels by factoring in "externalities" – the full financial burden to society from the adverse effects of oil and coal extraction, pollution and extreme weather. They cited the enormous benefits of removing these subsidies:

> Eliminating post-tax subsidies in 2015 could raise government revenue by $2.9 trillion (3.6% of global GDP), cut global CO2 emissions by more than 20% and cut premature air pollution deaths by more than half.[4]

Fossil fuel subsidies are well known to be harmful and inefficient. They

- make uneconomical dirty fuels economical, while increasing pollution and undermining public health
- act as a negative carbon tax, distorting the playing field for energy efficiency, and starving innovation in renewables
- create a market advantage for fossil fuel producers over clean energy producers
- create "zombie energy" projects that would be economically unviable without government support
- displace funds for basic needs such as education, health care, and environmental protection
- lock in fossil fuel dependency by giving strong signals to investment decision makers
- maintain an addiction to cheap energy that society feels it cannot do without.

The fast-reacting power of the market would move investment money out of fossil fuel energy and into renewable energy. Removing fossil fuel subsidies would

- increase production costs and decrease the fossil fuel supply
- increase prices, thereby lowering consumption and emissions
- encourage greater energy efficiency through the substitution of alternative energy, thus reducing CO2 emissions
- generate low-carbon innovation while boosting economic growth.[5]

In 2009, recognizing their inefficiency, the G20 nations pledged to phase out fossil fuel subsidies. Since then, while India and China and many developing nations have been steadily reducing subsidies,[6] the US and Canada have stalled.

China's annual fossil fuel subsidies were reported by the IMF in 2015 as $1,844 billion US, as compared to US. subsidies of $605.9 billion. China's emissions are twice those of the United States, and its fossil fuel subsidies are three times.

As predicted long ago, China's own poor and the poor in all countries are being hit first and hardest by global climate change. The IMF, World Bank, OECD and the International Energy Agency have all reported that fossil fuel subsidies are harmful to economies and societies as well as to the environment. A 2013 report by the Overseas Development Institute summed it up:

> Fossil fuel subsidies undermine international efforts to avert dangerous climate change and represent a drain on national budgets. They also fail in one of their core objectives: to benefit the poorest. Phasing out fossil fuel subsidies would create a win-win scenario. It would eliminate the perverse incentives that drive up carbon emissions, create price signals for investment in a low carbon transition and reduce pressure on public finances.[7]

The backing of oil by government has been described by the International Institute for Sustainable Development as a "confidence trick," encouraging private investment in fossil fuels.[8] It should be recognized as illegal in international law and placed under the auspices of a monitoring body, perhaps within the United Nations. China and the U.S. in particular must be required to halt their enormous subsidies immediately.

Tax Reform to Mitigate Catastrophic Climate Change

Maintaining fossil fuel subsidies negates the urgent statement of incoming UN Secretary-General António Guterres in May 2017:

> It is absolutely essential that the world implements the Paris Agreement – and that we fulfill that duty with increased ambition ... As the Intergovernmental Panel on Climate Change put it: "Human influence on the climate system is clear. The more we disrupt our climate, the more we risk severe, pervasive and irreversible impacts."[9]

The Costs to Society of Burning Carbon

Many regions have already had catastrophic extreme weather impacts, driven by climate change with single events affecting millions of people. A 2015 Oxfam assessment estimates the poorest half of the global population are responsible for approximately 10% of global emissions and yet live overwhelmingly in the countries most vulnerable to climate change – while the richest 10% of people in the world are responsible for about 50% of global emissions. "The richest 1% may emit 30 times more than the poorest 50%, and 175 times more than the poorest 10%."[10]

The 2006 UK Stern Commission Review, "The Economics of Climate Change," called for urgent measures, explaining in great detail the crippling catastrophic effect of unmitigated global climate change on the world economy. The Review sees climate change as the greatest and widest-ranging market failure in history. Market failures occur when the allocation of goods and services are fatally inefficient. In this case the oil industry and governments have promoted the negative externalities of GHG pollution, which include the health and clean-up costs imposed on the whole body of society. By avoiding the cost of this self-serving pollution they have made climate change the ultimate market failure.

The Stern Commission's remedy was first to measure the cost to society of a ton of carbon dioxide. They estimated that the 2007 social cost of carbon was $310 U.S. per ton.[11]

The 2007 IPCC 4th assessment estimated the costs of burning carbon as up to $350 U.S. per ton, and the social cost of greenhouse gases as rising over time at the rate of 2% to 4% per year.

A carbon pricing system was needed to pay for these harms while reducing the use of carbon and restoring market efficiency.

Carbon Pricing

Pricing carbon is levying a fee on CO2 emissions to offset the cost of the damage they cause. The price is the amount that must be paid for emitting one ton of CO2 into the air.

A carbon tax makes it financially preferable for all concerned to take steps to reduce emissions by scaling back production, changing production methods, or introducing abatement measures that cost less than the tax. (A carbon tax does not directly regulate emission volumes.)

Alternatively, under a cap-and-trade system, the government puts a cap on emission volumes through a prescribed number of emission allowances each year. These allowances can be auctioned to the highest bidder as well as traded on secondary markets, creating a carbon price determined by the market.

Cap-and-trade is a subterfuge designed to promote fossil fuel viability, allowing business-as-usual to continue. Caps on total national emissions can never be fair when per capita emissions from wealthy countries such as the US and Canada are approximately 20 times higher than those from countries such as India.

The only effective way to control emissions globally is for the wealthy nations of the Global North to implement an internal national carbon tax on themselves, in combination with placing a port-of-entry carbon tax on goods arriving from countries who have no similar national carbon tax. This import tax will be a strong incentive for other countries to develop parallel carbon tax systems, thus bringing about global participation.

Discussion has gone on for years over which method is best: carbon tax, cap and trade, or direct regulation. The European Union has had carbon trading since 2005. The EU carbon trade price was running at about $6 U.S. a ton in 2017, which is paltry in terms of social costs and gives permission to keep polluting. It is not good news that China and California have selected carbon trading as their pricing method.

Most economists favor a carbon tax over cap-and-trade. A carbon tax has been acceptable even to Exxon for some years but most carbon tax proposals have been, in terms of the magnitude of the problem, unrealistic political compromises.

What is needed is a calculated, fully-costed global pollution charge that treats all jurisdictions alike – rather than a negotiated and compromised fee that the fossil fuel industry can live with while maintaining business-as-usual.

This idea is not new. Governments agreed long ago on the polluter pays principle, on full costing of CO2, and on halting the externalization costs – under Agenda 21 at the 1992 UN Earth Summit. They reaffirmed these agreements (specifically Agenda 21) at the Rio+20 UN conference in 2012.

In 2009, Canada's National Round Table on Environment and Economy issued a study suggesting that economy-wide carbon pricing will need to rise to $108 per ton of CO2 equivalent by 2020 and upwards of $300 per ton of CO2-equivalent by 2050 in order to drive the social behavior and the technology needed to achieve the necessary deep reductions.

Lester Brown's 2009 "Plan B 4.0: Mobilizing To Save Civilization" included a carbon tax rising to $200 by 2020.

A 2015 Stanford study, "Temperature impacts on economic growth warrant stringent mitigation policy," published in *Nature Climate Change,* estimated that the social cost of carbon is about $220 per ton.

In practice, Finland became the first country to introduce carbon pricing in 1990; then came Sweden in 1991, where the carbon tax is 1100 SEK or $165 U.S. per ton of CO_2.

The IPCC 5th Assessment reported in 2014 that only the Scandinavian countries, the Netherlands, the UK, and the Canadian province of British Columbia had significant carbon taxes of at least $10 U.S. per ton. Although other nations, cities and regions have since followed suit, the World Bank reported that as of 2015, carbon pricing still only applied to 13% of GHG emissions worldwide.[12]

Benefits of a Carbon Tax

A well-designed carbon tax offers many and varied efficiencies:

- efficient tax administration if emissions from all sectors (agriculture, manufacturing, home heating, gasoline and diesel, and oil and gas extraction) are taxed at the same rate per ton of CO_2 equivalents;
- efficient administration if the tax is collected where fossil fuels enter the economy, such as ports, oil refineries, natural gas providers, and coal-processing plants;
- efficiency in encouraging investors and entrepreneurs to develop new low-carbon technologies;
- efficiency in offering stable carbon prices, so energy producers and entrepreneurs can make investment decisions without fear of fluctuating regulatory costs; and

- efficiency in generating substantial revenue, which can be returned to the public through dividends, or used to finance other tax reforms, or public transportation, walkways and bikeways.

Climate change policy in the United States (or any country) would be most effective if it were federal, economy-wide, and market-based. A carbon tax meets all these criteria. A tax that starts at a set rate and increases predictably over time would establish incentives throughout the economy to reduce carbon dioxide emissions with minimal disruption. And by encouraging a less carbon-intensive economy, a carbon tax would improve long-term energy security.

Public Acceptance of a Carbon Tax

To address the issue of public acceptance, we cite parts of a letter written by Dr. James Hansen to President-Elect Barack Obama in 2008:

> A "carbon tax with 100 percent dividend" is needed to reverse the growth of atmospheric CO2...The entire tax should be returned to the public, equal shares on a per capita basis...

> A tax should be called a tax. The public can understand this and will accept a tax if it is clearly explained and if 100 percent of the money is returned to the public...

> The public will take steps to reduce their emissions because they will be continually reminded of the matter by the monthly dividend and by rising fuel costs...

> Tax and dividend is progressive. A person with several large cars and a large house will have

a tax greatly exceeding the dividend. A family reducing its carbon footprint to less than average will make money...A carbon tax is honest, clear and effective. It will increase carbon prices but low and middle income people, especially, will find ways to reduce carbon emissions so as to come out ahead.[13]

The Citizens' Climate Lobby (CCL) proposes a revenue-neutral carbon fee and dividend tax on CO2 equivalents at the point upstream where they enter the economy – whether the well, mine, or port of entry. The fee rises annually from a starting point of $10 or $15/ ton per year by $10/ton per year. The upstream emissions model is simpler to apply and monitor than a downstream model, and covers all products and services that have a carbon footprint.

Under this model, 100% of the net fees are held in a trust fund and returned directly to households as a monthly dividend. About two-thirds of households would break even or receive more than they would pay in higher prices. This stimulates both the economy and a switch to low carbon products, while giving industry a known timeline and cost projection to restructure their business models.[14]

In spite of CCL and similar models, G20 public finance for *new* petroleum exploration proceeds unabated, averaging $13.5 billion a year between 2013 and 2015.[15]

These G20 policies persist in spite of market jitters over "stranded assets." Word is out that to avoid the worst impacts of climate change, even *known* oil reserves are "unburnable" and likely to lose value as investments.

But to date, policy proposals for carbon trading and carbon taxation fail to include termination of GHG subsidies, making them grossly inadequate as an emergency zero-emissions response.

This dogged retention of fossil fuel subsidies in spite of the G20 agreement in 2009 to phase them out might be explained by the U.S. reliance on the Petrodollar system. Launched in 1973 by Richard Nixon, the system requires OPEC signatories to accept US dollars for oil payment and nothing else.

If petroleum corporations lose their subsidies, become unprofitable, and are phased out, the demise of the Petrodollar will almost certainly follow. In early October 2017, Saudi Arabia, which wants to launch a major nuclear power program, joined with Russia in a $1 billion energy fund that will include cooperation in nuclear power.[15] The Trump administration, which has embraced the vestigial age of coal, would do well to see the writing on the wall: that most other countries are preparing for the post-petroleum order.

Other Taxes and Additional Measures

Other product taxes may be levied or credited to reduce emissions, including:

- a GHG-weighted tax applied to large farm animal food, and/or to the consumption of ruminant meat;
- a nitrogen fertilizer tax to curb nitrous oxide (N_2O) emissions from industrial fertilizers – a powerful GHG making up about 8% of society's GHG emissions.[17]

Other financial incentives include:

- transferring large-farm subsidies to small, bio-diverse, ecological farms around the world, given that agriculture is responsible for 25-30% of global GHG emissions, and following the Norwegian example[18] (see Chapter 9).
- tax deductible renewable energy investments; many jurisdictions have tax credits already. The US has had an investment tax credit for solar energy since 2005, allowing a 30% tax write-off

on costs. This should be raised to 100%. Instead, the U.S. announced in 2017 that this credit will be phased down to 22% by its final year, 2021.[19]

- green bonds, which raise investment for clean renewables. According to a report from Moody's rating agency, green bond issues were expected to surge to more than $200 billion in 2017. Green bonds have helped countries, cities and municipalities around the world raise much-needed finance at scale for investments in low-carbon transport and clean energy.

Conclusion

Government subsidies keep the price of oil artificially low. Unless there is carbon taxation, the price at the pump does not include the cost of damages that fossil fuels do to public health and the environment. However, a carbon tax cannot be efficient while it is offset by fossil fuel subsidies, which essentially act as a negative carbon tax.

The subsidy debate and the carbon pricing debate were shown to be two sides of the same coin in the 2015 IMF working paper, which suggested that:

> carbon prices and carbon taxes [should be added] into the price of the fuel, so that you're paying for the external environmental damages and so on within the price of that fuel. And if that's not priced in then that's counted as a subsidy. That's really changing the game in terms of subsidy definition.[20]

Current fossil fuel subsidy policy is not only contradictory and irrational but has predictably poor outcomes. A team of economists and climate scientists reported in 2017 that "Unmitigated climate change will make the United States poorer and more unequal ... the poorest third of counties could sustain economic damages costing as much as 20 percent of their income if warming proceeds unabated."[21]

In summary, according to the IMF:

> Fiscal instruments (carbon taxes or similar) are the most effective policies for reflecting environmental costs in energy prices and promoting development of cleaner technologies, while also providing a valuable source of revenue (including, not least, for lowering other tax burdens) ... Getting energy prices right has large fiscal, environmental, and health benefits at the national level, and need not wait for international action.[22]

Unless governments step up to the plate to remove subsidies and implement carbon taxation, the atmosphere will continue to serve as a free dumping ground for the criminal release of catastrophic heat-trapping GHG emissions.

Endnotes

1   Oil Change International, *Fossil Fuel Subsidies Overview,* 2017.
2   International Energy Agency, *Fossil Fuel Subsidy Database*, OECD/IEA, 2017 (http://www.worldenergyoutlook.org/resources/energysubsidiesfossilfuelsubsidydatabase).
3   David Coady et al., "How Large are Global Energy Subsidies?" IMF, 2015 (http://www.imf.org/en/Publications/WP/Issues/2016/12/31/How-Large-Are-Global-Energy-Subsidies-42940).
4   Ibid., 6.
5   International Institute for Sustainable Development, "Zombie Energy: Climate benefits of ending subsidies to fossil fuel production," IISD, February 2017, iv (http://www.iisd.org/sites/default/files/publications/zombie-energy-climate-benefits-ending-subsidies-fossil-fuel-production.pdf).
6   International Energy Agency, *Energy Subsidies by Country, 2015 (including online database)*, OECD/IEA, 2017.
7   Shelagh Whitley, "Time to Change the Game: Fossil Fuel Subsidies and Climate," Overseas Development Institute, November 2013.
8   "Zombie Energy," iv.

9   United Nations News Centre, "Climate action 'a necessity and an opportunity,' says UN chief, urging world to rally behind Paris accord," 30 May 2017 (http://www.un.org/apps/news/story.asp?NewsID=56865#.WUwhh1EXT1J).

10   Oxfam, "Oxfam Media Briefing: Extreme Carbon Inequality," 2 December 2015.

11   United Kingdom, Stern Commission, *The Economics of Climate Change*, Cambridge University Press, 2007.

12   World Bank Group, "State and Trends of Carbon Pricing," September 2015 (http://documents.worldbank.org/curated/en/636161467995665933/pdf/99533-REVISED-PUB-P153405-Box393205B.pdf).

13   James E. Hansen, "Tell Barack Obama the Truth – The Whole Truth," 2008, p. 4

14   Citizens' Climate Lobby, "How a Carbon Fee and Dividend Works," 2017.

15   Alex Doukas et al., "Talk is Cheap: How G20 Governments are Financing Climate Disaster," Oil Change International, 2017.

16   "Russia & Saudi Arabia sign billion dollar deals during King's visit," *RT*, 5 October 2017.

17   IPCC 4th Assessment, "Climate Change 2007 Synthesis Report," 2007.

18   Dr. Mae-Wan Ho, "Paradigm Shift Urgently Needed in Agriculture: UN Agencies Call for an End to Industrial Agriculture Food System," Permaculture Research Institute, 18 September 2013.

19   Ashlea Ebeling, "Tax Credits for Going Green: Grab Them Before It's Too Late," *Forbes*, 21 April 2017.

20   CarbonBrief, "Explainer: The challenge of defining fossil fuel subsidies," 12 June 2017.

21   Phys.Org, "Study: Climate change damages US economy, increases inequality," 29 June 2017 (https://phys.org/news/2017-06-climate-economy-inequality.html).

22   International Monetary Fund, "Climate, Environment, and the IMF," 17 April 2017 (http://www.imf.org/en/About/Factsheets/Climate-Environment-and-the-IMF).

Chapter Eight

HUMAN RIGHTS-BASED LEGAL CHALLENGES

Professor Lawrence Torcello of the Rochester Institute of Technology in New York specializes in moral and political philosophy. In April 2017 he wrote that "there can be no greater crime against humanity than the foreseeable and methodical destruction of conditions that make human life possible," adding that

> If any nation were to enact policies calculated to systematically destroy cultural lands and displace native people, as climate change will, it would rightly raise international debates over genocide. It makes no difference to populations forced off their homelands whether the resource exploitation responsible is occurring in West Virginia or Papua New Guinea.
>
> The climate policies of the Trump administration, backed by many Republican leaders, are rooted in culpable ignorance and transparent corruption. And they place us all at risk on a scale that previous crimes against humanity never have.[1]

This chapter addresses the questions: Why is the destruction of the life-sustaining biosphere allowed to continue, and at an increasing rate, to this day? How can this be legal? Common sense and instinctive morality says it cannot be.

Recap: The Scope and Urgency of the Moral Problem

The world has been informed for 25 years by UN reports, intergovernmental reports, and environmental NGO reports that the economics of the wealthy and wasteful consumer culture of the Global North – now being rapidly globalized – is not sustainable.

The latest data and reports show that the natural world is being polluted, degraded and destroyed so rapidly that global land and ocean ecosystems are collapsing, with species being eliminated at a rate of 1,000 times the pre-human extinction rate.

At the same time, the human rights of people over much of the world are being sacrificed to rising seas, extreme weather events, and loss of crops.

At this juncture it is clear that our systems of economics and government are miserably failing to address this planetary emergency. As past chapters have shown, this reflects an unprecedented collapse of ethics and morality led by the Global North. Both Pope Francis, head of the Roman Catholic Church, and his counterpart, Patriarch Bartholomew, leader of the Eastern Orthodox Catholic Church, have declared that it is a sin against creation for human beings to change earth's climate.

As Ronald Dworkin famously said, "Moral principle is the foundation of law." When humanity's natural sense of moral responsibility collapses, ethical behavior must be restored by enforcing the law that represents the moral principles of society.

Applying Legal Action to Ethical Failure

As we have seen, it is on the record that multinational fossil fuel corporations such as ExxonMobil have for decades connived and lied to policymakers and to the public to forestall protective measures that would control greenhouse gas emissions.

It has long been the central duty of governments to protect citizens of the state from such treachery. (The ancient Public Trust Doctrine, to be discussed below, reflects that historic tradition.)

We have seen that the United States and China are the two

key nations needed to lead action to address the climate crisis. China is rising to the crisis, but what about the U.S.? With the continuing failure of the U.S. government to protect its citizens (and the world) from U.S. emissions, the most important legal case today may be *Our Children's Trust*. This action (which will also be discussed below) is seeking to hold the U.S. government legally accountable for failing to protect the climate that its children's future lives and livelihoods depend upon.

A Brief History of Failed International Attempts to Curb Climate Change

International environmental law really got started with the Stockholm Declaration of the United Nations Conference on the Human Environment, adopted June 16, 1972.

The Stockholm Declaration states that "the natural resources of the earth, including the air, water, land, flora and fauna, and especially representative samples of natural systems, must be safeguarded for the benefit of present and future generations." Principle 6 referred, at this early date, to "the release of heat":

> The discharge of toxic substances or of other substances and the release of heat, in such quantities or concentrations as to exceed the capacity of the environment to render them harmless, must be halted in order to ensure that serious or irreversible damage is not inflicted upon ecosystems. The just struggle of the peoples of all countries against pollution should be supported.[2]

The UN Framework Convention on Climate Change (UNFCCC): The UNFCCC is the best known international legal protection for the climate. This agreement grew out of UN General Assembly resolutions and negotiations from 1988 to 1990, was entered into by all nations in 1992, and is binding to this day. No nations have withdrawn.

The UNFCCC referred back to the provisions of both the 1972 Stockholm Declaration and UN General Assembly Resolution A/RES/44/207 of 22 December 1989, which called for the protection of the global climate for present and future generations.

The purpose of the 1992 Convention is clear in the Preamble and throughout the text.

Article 2, Objective:

> The ultimate objective of this Convention and any related legal instruments that the Conference of the Parties may adopt is to achieve in accordance with the relevant provisions of the Convention, is the stabilization of greenhouse gas concentrations in the atmosphere at a level that would prevent dangerous anthropogenic interference with the climate system. Such a level should be achieved within a time frame sufficient to allow ecosystems to adapt naturally to climate change, to ensure that food production is not threatened and to enable economic development to proceed in a sustainable manner.[3]

Article 3 includes the precautionary principle:

> 3. The parties should take precautionary measures to anticipate, prevent, or minimize the causes of climate change and mitigate its adverse effects.[4]

Under Article 4, Commitments:

> 1(b)   All parties shall formulate, implement, publish and regularly update national and, where appropriate, regional programmes containing measures to mitigate climate change by addressing anthropogenic emissions.[5]

There are also clear and specific requirements for the industrialized high-emitting countries, which include obligations to the most vulnerable and least developed low-emitting countries.[6] Other UN Conventions are linked to the UNFCCC, giving it added weight. These include the 1979 Convention on Long-Range Transboundary Air Pollution, the 1992 Convention on Biological Diversity, and the 1994 Convention to Combat Desertification.

Central to avoiding planetary catastrophe is respect for the UN Climate Convention. However, the Convention is poorly understood with regard to the danger of climate change, which is not being sufficiently respected or reported.

Dangerous climate change is clearly and specifically defined by the UNFCCC in terms of dangerous interference with the climate system. Many people do not realize that danger was defined by the Convention in 1992 as unsafe levels of atmospheric GHGs. Safety is defined as a level "to allow ecosystems to adapt naturally to climate change, to ensure that food production is not threatened and to enable economic development to proceed in a sustainable manner."[7]

The Kyoto Protocol was adopted in 1997 and entered into force in 2006. In March 2001, soon after coming to power, Republican President George W. Bush reneged on this Protocol, which as a vital extension to the UNFCCC, committed signatories to reducing GHG emissions to 7% below those of 1990 by the year 2012. Effectively sabotaging the Protocol, Bush blamed incomplete scientific knowledge on global warming, and the lower emissions reduction requirements of developing countries.

The Paris Agreement: Like the Kyoto Protocol, the 2015 Paris Agreement was supposed to implement the 1992 Climate Convention in a legally effective manner. However, it does not – among other reasons because there are no legally binding limits to emissions, atmospheric greenhouse gas concentrations, or degrees of global warming. This was the case even though, as the Paris Conference began, GHG emissions were at an all-time high, atmospheric GHG concentrations were at an all-time high with CO2 accelerating, and global warming was at an all-time high. This acceleration continued into 2016 and 2017 (see Science Appendix).

Nonetheless, the Paris Agreement achieved enough ratifications (55%) to enter into force in November 2016. As of December 2017, 171 out of 197 signatories had ratified. However, once again the US reneged when the Republican Trump administration issued notice of withdrawal in June 2017.

The US, now an oil exporter, will by reneging join the list of oil producers that have not ratified, the largest being the Russian Federation. Other significant oil and gas producers that have not ratified include Iran, Kuwait, Iraq, Venezuela, Oman, Uzbekistan, and Ecuador.

Failed Regulation of Greenhouse Gases in the United States

The United States *Clean Air Act*, originally passed in 1970 and updated in 1990, is the comprehensive federal law that regulates air emissions. Among other things this law authorizes the Environmental Protection Agency (EPA) to establish National Ambient Air Quality Standards to protect public health and to regulate emissions of hazardous air pollutants.

The most significant legal finding on greenhouse gas dangers in the United States is the 2009 U.S. EPA Endangerment Finding.[8] This comprehensive scientific and legal assessment, which received more than 380,000 public comments, took years to complete and was signed into law on December 7, 2009 under the *Clean Air Act*.

This assessment found that the current and projected concentrations of the six key well-mixed greenhouse gases in the atmosphere – carbon dioxide (CO_2), methane (CH_4), nitrous oxide (N_2O), hydrofluorocarbons (HFCs), perfluorocarbons (PFCs), and sulfur hexafluoride (SF_6) – "endanger both the public health and public welfare of current and future generations."[9]

The long technical Endangerment Finding repeatedly describes atmospheric greenhouse gas emissions as pollution. The obvious fact that these emissions are a form of pollution is central to the problem.

The United States, like other developed countries, has a

process of scientific pollution risk assessment and regulation for the public safety. An independent science committee conducts a risk assessment of the pollutant in question and then determines a safe environmental level for the most vulnerable populations (often small children).

There is no government involvement in this process. The committee examines input from experts and from the public and then records its conclusions and recommendations. Potentially deadly pollutants may not be approved by the committee at all, or they may be approved at an extremely small concentration.

The report is then presented to the government environmental agency and published for review by concerned stakeholders and the public. For a high-risk pollutant or in a contentious situation the government agency may hold a public review to hear formal presentations of evidence. Finally, the agency puts the whole report together and gives it to the government to decide on what action to take.

In practice this process is often a long fight between polluting corporations and public interest organizations. Although in 2010 the EPA adopted greenhouse gas emission standards for new cars and light-duty vehicles, there has been no across-the-board regulation of GHG emissions in the United States.[10]

It is extraordinary that GHGs have escaped the standard risk assessment orders that regularly control other hazardous substances such as carcinogens and pesticides. It would seem that the atmosphere has somehow fallen through the cracks of pollution control. This may be because the climate change assessments of the Intergovernmental Panel on Climate Change are regarded as fulfilling this role. However the IPCC does not follow the same process. It makes no conclusions or recommendations on dangerous pollutants, and has no authority to regulate them.

It is clear that another approach is needed.

The Public Trust Doctrine (PTD)

The ancient Public Trust Doctrine requires government stewardship of "the common wealth" – the natural resources upon which society depends for the benefit of existing and future generations. It is strengthened by its long and continuous European history extending as far back as sixth-century Rome.

In general, a trust's assets (its *res*) are owned by two parties: the beneficiaries, who have beneficial ownership of the assets, and the trustees, who are the legal owners. As legal owners, the trustees have control of the assets, but they have this control entirely for the sake of the beneficiaries.[11]

The essence of a trust is a fiduciary relationship that imposes on trustees a duty to act for the benefit of beneficiaries with respect to trust matters.

In the Public Trust Doctrine, government acts as a trustee, with the management responsibility and accountability similar to that of oversight of an estate or investment account. In the United States the PTD was first applied to coastlines and stream beds but has since expanded to include the guarantee of clean water and clean air.

The idea of the public trust is manifest in the legal systems of many nations throughout the world. Professor Charles Wilkinson has traced the doctrine to the ancient societies of Europe, Asia, Africa, Muslim countries, and to Native America, where stewardship of nature has been central to indigenous governance from time immemorial.[12]

The Public Trust Doctrine as a Supplement to the UN Human Rights Treaty System[13]

As we have seen, atmospheric GHG emissions heat the lower atmosphere, which is the air around and above us. Although in 2009 the EPA ruled that these emissions are hazardous atmospheric pollutants, the US Republican Party has managed for years to prevent regulatory measures from being imposed. This continued in March 2017 when President Trump signed an executive order to block Obama's Clean Power Plan restricting GHG emissions from coal-fired power plants.

Following Earth Day in 1970, the formation of the IPCC in 1988, and the 1992 U.N. Earth Summit at Rio – at which the nations signed the U.N. Framework Convention on Climate Change (UNFCCC) – there was great optimism that global warming could be stopped before it became much worse. But the world's nations have failed to use the UNFCCC to stop or even slow climate change.

Some people from the beginning feared that this approach to the problem would fail, because it is a treaty system in which nations can veto any proposals they might find too costly or otherwise undesirable. Accordingly, any treaties achieved are usually weakened by the effort to achieve consensus, are unable to prevent much pollution, and are difficult to force upon sovereign nations. Any country can drop out of the treaty at will, as has happened with the Kyoto Protocol.

A supplement to the treaty system would be to employ the PTD internationally. This would not be as problematic because the principles of public trust can be easily understood and many countries have already developed the PTD at the national level.

A common legal basis is provided by the UNFCCC formula of "common but differentiated responsibilities," which means that all nations have responsibilities to reduce fossil fuel emissions, but high-emitters need to reduce their emissions more quickly.

In September 2017, a review by the European Court of Auditors warned that to meet 2030 emissions targets the EU will need to increase annual emissions reductions by 50%. To achieve this, the required implementation by individual member states could be furthered by adopting an integrated public trust approach.

As we have seen, America and other industrial nations have resisted the UNFCCC formula of "common but differentiated responsibilities," using their objection as an excuse to do little about the climate. Within the treaty framework, there is nothing the poorer nations can do about this.

To solve this problem, the international PTD movement is counting on the domestic judiciaries to play their role. Prof. Mary Christina Wood of the University of Oregon law school explains:

As a legal doctrine, the public trust compels protection of those ecological assets necessary for public survival and community welfare. The judicial role is to compel the political branches to meet their fiduciary obligation through whatever measures and policies they choose, as long as such measures sufficiently reduce carbon emissions within the required time frame.[14]

That describes the function of the PDT within the nation.

The hope is that the judiciaries around the world will do this in their own countries as a support to the international treaty system – especially with regard to Atmospheric Trust Litigation, which will be discussed below.

Nature's Planetary Trust

In describing "nature's trust," Professor Wood wrote:

The *res* of Nature's Trust consists of ecological assets, natural wealth that must sustain all foreseeable future generations of humanity. It amounts to humanity's survival account – the only one it has. Government trustees must protect trust resources for the benefit of present and future generations.[15]

Looking at government as trustees to preserve natural resources for their rightful beneficiaries – including the people of future generations – would imply policies and behaviors radically different from those that have brought us to the brink of destruction.

That government officials should act as nature's trustees has a long history. "This idea needs to be understood not as a new one," writes Professor Peter Brown, but as "the rediscovery of an old one that has been lost to political consciousness."[16]

The recovery of this idea began in the latter part of the 20th century with Joseph Sax's 1970 article, "The Public Trust Doctrine in Natural Resource Law: Effective Judicial Intervention." This article was written in response to the desire of lawyers and other citizens to have a legal basis for the rights of the general public – a basis that could be enforceable against the government. This basis, Sax suggested, could best be provided by the Public Trust Doctrine.

Some U.S. states, especially California, Hawaii, and Pennsylvania, and some countries, especially India, Kenya, and South Africa, have demonstrated the potential of the PTD for preserving and restoring various natural resources.[17] The crucial question at this point is whether it can provide a legal basis for preventing climate collapse.

Atmospheric Trust Litigation (ATL)

The Atmospheric Trust Litigation attempts to do this. It "simply applies the public trust doctrine to the atmosphere," says Professor Wood. This doctrine concerns "resources that the public relies on for its very survival," and the "atmosphere certainly qualifies."[18]

Atmospheric Trust Litigation is simply an extension of a familiar type of suit, in which the court is asked to defend the "commons" against private companies encroaching on it. It differs only in defending the Earth's largest commons: the atmosphere, considered as a commonly-shared public trust resource.

This extension of the public trust doctrine is legitimate, argue defenders, because the doctrine deals with matters of "public concern to the whole people" (as the Supreme Court put it in 1892), and there is no more obvious resource of public concern today than "the atmosphere upon which all life depends."[19]

"By linking to scientific prescriptions as the measure of fiduciary responsibility," Wood says, "the ATL approach is aimed at divesting the world's political leaders of their assumed prerogative to take action only according to their political objectives."[20]

ATL focuses on the atmosphere as "a single asset in its entirety" and "characterizes the United States as a trustee, and each of the 50 states as co-trustees, of the atmosphere." It also "characterizes all nations on Earth as co-tenant sovereign trustees of that asset."[21]

ATL refers to suits filed against governments, asking the judicial branch of government to find the executive branch and some of its agencies guilty of a breach of trust with regard to their responsibility to safeguard the atmosphere, and to force them to stop the breach-causing behavior.

By not being responsible trustees of the atmosphere, the government is failing to preserve an essential part of the trust's life-giving endowment. Failing to preserve the trust – by allowing fossil fuel companies to spew their waste products into the atmosphere – amounts to robbing today's children. The atmosphere also belongs to future generations, so that "failure to safeguard it amounts to generational theft."[22]

Politicians have gotten away with this theft because children, and future generations in general, do not vote. Children need to rely on our politicians, policed by our courts, to be faithful trustees of the natural resources that they will need.

Our Children's Trust Lawsuit

In 2011, *Juliana v. United States et al.* was filed in Federal Court by 21 young plaintiffs who argue that their constitutional and public trust rights are being violated by the U.S. government.

World-renowned climate scientist Dr. James E. Hansen is acting as co-plaintiff on behalf of his granddaughter. Our Children's Trust, which leads the lawsuit, advocates on behalf of youth and future generations, and towards legally-binding science-based climate recovery policies.

The lawsuit asserts that young people are guaranteed a non-toxic environment fit for human habitation as part of their constitutional rights to "life, liberty, and the pursuit of happiness." The case is predicated on meeting the intent of the 2015 Paris climate agreement to keep global average temperatures at or below 1.5° Celsius.

According to court documents,

> [The] Plaintiffs in this civil rights action are a group of young people between the ages of eight and nineteen ("youth plaintiffs"); Earth Guardians, an association of young environmental activists; and Dr. James Hansen, acting as guardian for future generations.
>
> Plaintiffs filed this action against defendants the United States, President Barack Obama, and numerous executive agencies.
>
> The plaintiffs allege that the defendants have known for more than fifty years that the carbon dioxide produced by burning fossil fuels was destabilizing the climate system in a way that would "significantly endanger plaintiffs, with the damage persisting for millenia." Despite that knowledge, plaintiffs assert, defendants, "[b]y their exercise of sovereign authority over our country's atmosphere and fossil fuel resources, ... permitted, encouraged, and otherwise enabled continued exploitation, production, and combustion of fossil fuels, ... deliberately allow[ing] atmospheric CO2 concentrations to escalate to levels unprecedented in human history. [23]

Children have standing under the public trust doctrine to bring a climate action lawsuit. Specifically the children are seeking a court-issued and court-monitored order over Executive Branch officials and agencies. They seek "an enforceable national remedial plan to phase out fossil fuel emissions and draw down excess atmospheric carbon dioxide" and "to restore Earth's energy balance."[24]

This plan would require the federal government "to cease their permitting, authorizing, and subsidizing of fossil fuels and, instead, move to swiftly phase out carbon dioxide emissions, as well as take such other action as necessary to ensure that atmospheric carbon dioxide is no more concentrated than 350 ppm by 2100."[25] This is what the science dictates, as covered in the science appendix to the case.

It's a David and Goliath match. The plaintiffs are up against every big US-based oil corporation. Canadian author Naomi Klein and Bill McKibben of 350.org have called this "the most important lawsuit on the planet right now."

Many others agree. In 2013, seven *amici curiae* (friends of the court) briefs were filed, showing broad-based support for the young people. These included the National Congress of American Indians, prominent scientists and legal scholars, faith and human rights groups, and national security experts, who stated that action by government to reduce GHGs is essential to mitigate the worst security consequences of climate change.

Concern from those implicated gradually mounted. On January 13, 2016, the District Federal Court in Oregon granted intervenor status (an intervenor is a third party who at the discretion of the Court intervenes in a legal proceeding) to this list of applicants representing nearly all the world's fossil fuel companies: the American Fuel and Petrochemical Manufacturers (AFPM) representing Exxon Mobil, BP, Shell, Koch Industries, and virtually all other U.S. refiners and petrochemical manufacturers; the American Petroleum Institute (representing 625 oil and natural gas companies), and the National Association of Manufacturers. They then filed to have the case dismissed.[26]

Immediately after this, the Center for Earth Jurisprudence, on behalf of the Global Catholic Climate Movement, a network of 250 Catholic organizations, and the Leadership Council of Women Religious filed an *amicus curiae* brief in support of the children. The Catholic groups filed their briefs to make their views known that the youth's legal claims are rooted in U.S. traditions and parallel Roman Catholic tenets.

In an April 2016 landmark decision, Judge Coffin rejected the government and fossil fuel industry's Motions to Dismiss, and approved the constitutional climate case to go forward. In doing so, Judge Coffin characterized the case as an unprecedented lawsuit addressing "government action and inaction" resulting "in carbon pollution of the atmosphere, climate destabilization, and ocean acidification."[27] In his comments, Judge Coffin added:

> The intractability of the debates before Congress and state legislatures and the alleged valuing of short term economic interest despite the cost to human life, necessitates a need for the courts to evaluate the constitutional parameters of the action or inaction taken by the government. This is especially true when such harms have an alleged disparate impact on a discrete class of society.

In November 2016 the Court again rejected U.S. government and fossil fuel industry motions to dismiss the case, deciding in favor of the "groundbreaking" constitutional climate lawsuit, making clear that it would go quickly to trial.

In February 2017, counsel for the fossil fuel defendants said his clients did not admit CO2 levels had reached 400 ppm (when NOAA placed levels at 406.42 ppm).

Also in February 2017, the case was transferred from naming President Obama to naming Donald J. Trump as defendant.

In May 2017 the government and the industry attempted to derail the case with a rare early appeal, which was denied. Then, in an unusual procedural move, and after having fought hard to gain intervenor status the year before, the American Petroleum Institute, the American Fuel & Petrochemical Manufacturers, and the National Association of Manufacturers (NAM) filed motions to the Court requesting permission to withdraw from the lawsuit.

It was then reported that "NAM's motion to withdraw from the Our Children's Trust lawsuit came on May 22nd, just as it was about to be ordered to turn over documents on its climate change knowledge and activities, which would presumably have included its participation in political front and lobbying groups that denied the reality of climate change and spread disinformation on the subject."[28]

In June 2017, the Court issued an order releasing the fossil fuel industry defendants from the case, and setting a trial date for February 5, 2018 at the U.S. District Court of Oregon in Eugene. Youth plaintiffs, now age 10 to 21, and their attorneys began preparing for trial.

In a last-ditch effort to halt the case, the Trump Administration filed an extraordinary writ of mandamus petition with the Ninth Circuit Court of Appeals, seeking a stay of proceedings. Yale law professor Douglas Kysar, who is not connected with the case, wrote:

> Writs of mandamus are reserved for the most extraordinary and compelling situations in which ordinary rules of appellate procedure must be overridden to avoid a manifest injustice. For the Trump Justice Department to even seek a writ of mandamus in the current context is offensive to Judge Aiken, to the entire federal judiciary, and, indeed, to the rule of law itself. The writ should not be granted and we should all question why the Trump administration's lawyers are willing to try such a trick rather than forthrightly defend the case.[29]

In July 2017, the Ninth Circuit Court of Appeals ordered a temporary stay on the district court proceedings, and, in order for the opposing parties to present for and against the writ, ordered the child plaintiffs to answer it, which they did in late August.

Answering the writ, the children's attorneys made clear that the U.S. government had already admitted that its actions imperil youth plaintiffs with "dangerous, and unacceptable economic, social, and environmental risks," and that "the use of fossil fuels is a major source of [greenhouse gas] emissions, placing our nation on an increasingly costly, insecure, and environmentally dangerous path."[30]

Meanwhile the trial was pending before U.S. District Judge Ann Aiken in Eugene, Oregon, on February 5, 2018. The Children's Trust accepts tax-deductible donations[31] and needs strong support from all who are concerned about securing a livable, human rights-based climate.

Endnotes

1   Lawrence Torcello "Yes, I am a climate alarmist. Global warming is a crime against humanity," *The Guardian*, 29 April 2017.

2   United Nations, *Declaration of the United Nations Conference on the Human Environment*, 1972 (http://www.un-documents.net/unchedec.htm).

3   United Nations Framework Convention on Climate Change, 1992. (https://unfccc.int/resource/docs/convkp/conveng.pdf).

4   Ibid.

5   Ibid.

6   UNFCCC, *Guide to the Climate Change Negotiation Process*,1992 (http://unfccc.int/not_assigned/b/items/2555.php).

7   United Nations Framework Convention on Climate Change, 1992. Cited from Article 2, Objective.

8   The legal background for the *EPA's Endangerment Finding* is available at https://19january2017snapshot.epa.gov/sites/production/files/2016-08/documents/endangermentfinding_legalbasis.pdf.

9   Ibid.

10   Center for Climate and Energy Solutions, "EPA Greenhouse Gas Regulation FAQ," [2010] (https://www.c2es.org/federal/executive/epa/greenhouse-gas-regulation-faq).

11   Mary Christina Wood, *Nature's Trust: Environmental Law for a New Ecological Age*, Cambridge University Press, 2014, xiv.

12   As cited by Douglas Quirke, *The Pubic Trust Doctrine: A Primer*, Univesity of Oregon School of Law, Environmental Law and Natural Resources Law Center, 2016, 1.

13   Ved P. Nanda and William R. Rees, "The Public Trust Doctrine: A Viable Approach to International Environmental Protection," *Ecology Law Quarterly*, January 1976. The international role of the PTD is well established in the literature, as shown by the 194 footnotes to this scholarly paper (http://scholarship.law.berkeley.edu/cgi/viewcontent.cgi?article=1104&context=elq).

14   Mary Christina Wood, "Atmospheric Trust Litigation Across the World," In: Charles Sampford et al., *Fiduciary Duty and the Atmospheric Trust*, Routledge, 2012, 112.

15   Wood, *Nature's Trust*, 2014, 143.

16   Peter Brown, *Restoring the Public Trust: A Fresh Vision for Progressive Government in America*, Beacon Press, 1994, 69.

17   Michael C. Blumm & R.D. Guthrie, "Internationalizing the Public Trust Doctrine: Natural Law and Constitutional and Statutory Approaches to Fulfilling the Saxion Vision," *University of California at Davis Law Review*, 741, 745-46 (2012).

18   Fen Montaigne, "A Legal Call to Arms to Remedy Environment and Climate Ills," *Yale Environment 360*, 2 January 2014.

19   Michael C. Blumm and Mary Christina Wood, "Undermining Oregon's Public Trust Doctrine: Guest Opinion," *Oregonian,* 5 December 2013.

20   Wood, *Nature's Trust*, 127.

21   Wood, "Atmospheric Trust Litigation Across the World," 105.

22   Christina Wood, "Atmospheric Trust Litigation," in *Climate Change: A Reader*, ed. by W.H. Rodgers Jr. et al, Carolina Academic Press, 2011, 1034.

23   *Juliana v. United States, et al.*, No. 6:15-CV-01517-TC, 2016 WL 6661146 (D. Or. Nov. 10, 2016) (https://biotech.law.lsu.edu/cases/environment/Juliana.htm).

24   Ibid.

25   Ibid.

26   *Juliana vs. United States.* "Case Documents" (http://climatecasechart.com/case/juliana-v-united-states).

27   "Judge Coffin Rules in Favor of Youth Denying Motions to Dismiss," 8 April 2016 (https://www.ourchildrenstrust.org/federal-proceedings/)

28   Dan Zegart, "National Association of Manufacturers Attempts 11th Hour Escape from Our Children's Trust Climate Lawsuit," *DesmogBlog*, 23 May 2017.

29   Megan Darby, "Trump lawyers try 'extraordinary trick' to quash youth climate case," *Climate Change News*, 12 June 2017.

30   Our Children's Trust, "Details of Proceedings," 28 August 2017.

31   Donations may be made online at https://www.ourchildrenstrust.org/donate/

Chapter Nine

GAME CHANGERS IN TECHNOLOGY & INNOVATION

The 2014 IPCC assessment established a target of near-zero CO2-equivalent emissions (CO2, methane, nitrous oxide and halocarbons), which with today's technology could be met by 2050.

In order to meet this target, the people of the world need an immediate emergency response plan, prioritized by science-based effectiveness and rapid feasibility.[1]

This in turn requires the immediate declaration of a climate emergency by the world's governments, and the IPCC and UN Security Council.

The declaration of a global emergency gives national governments special constitutional powers to introduce emergency measures.

The following priority actions should be first on the agendas of all governments, intergovernmental organizations, institutions (such as national science academies), NGOs and corporations

1. Launch a WWII-type emergency mobilization to reduce emissions, such as has been proposed in depth by The Climate Mobilization.[2]
2. Implement a fully-costed escalating carbon pollution tax nationally and globally.

3. Terminate subsidies for fossil fuels by all governments in short order and transfer them to clean renewable energy sources.
4. In high-emitting countries, promote advanced safe nuclear fission reactors and small fusion reactors already under development (e.g., in the U.K., Canada and California). Fund small fusion reactors under development (e.g. in the US, UK, Canada and California).
5. Phase out subsidies for industrial chemical-intensive agriculture; switch subsidies to regenerative organic mixed agriculture, integrating woodlands.
6. Create global trusts in support of the Amazon and Boreal forests[3], perhaps taking lessons from the Antarctic regional governance model.[4]
7. Promote the value, variety and human health benefits of vegan and vegetarian food options;[5] otherwise the rate of deforestation (mainly for livestock) will increase, along with population growth and livestock methane emissions
8. Disincentivize the world fleet of one- or two-person gas-powered passenger vehicles, which is the largest source of road transportation CO_2 emissions. This mode of transport is on its way to becoming obsolete, being replaced by electric vehicles and a global network of charging stations, as well as by convenient, integrated, efficient and affordable public transit systems. (Tesla is developing a long-haul electric semi-truck that can drive itself and move in "platoons" that automatically follow a lead vehicle. These trucks are expected to get about 600 miles on a single charge without the need for diesel or drivers.)
9. Support the switch away from oil-based air travel, which must and can end, as will be discussed below

10. Require GHG pollution and mitigation education nationally and globally, supported by international agreement. The 2006 Stern Commission proposed a "Public education and persuasion" campaign, which could be started immediately. In December 2014, the UNFCCC urged, in the Lima Ministerial Declaration on Education and Awareness-raising, "all Parties to give increased attention, as appropriate, to education, training, public awareness, public participation and public access to information on climate change."

11. Increase to the maximum the natural carbon-storing capacity (uptake and retention time) of forests, vegetation and soils. This will require carbon pricing and public funding.

12. Take ocean acidification seriously. Direct air extraction of CO2, as well as zero combustion for energy, is essential to save the oceans. Acclaimed climate change writer Dr. Tim Flannery offers a nature-based way to reduce ocean acidification in his 2017 book, *Sunlight and Seaweed: An Argument for How to Feed, Power and Clean Up the World.*

13. Support the development of direct CO2 air removal technologies. According to physicist and climate scientist Bill Hare of Climate Analytics, the only thing now that can save coral reefs, biodiversity and food production is negative emissions technology.[6]

Carbon Engineering is a Canadian company that has raised about $40 million and extracts approximately one tonne of carbon dioxide each day with turbines and filters.

Climeworks is a Swiss company that opened in July 2017, hoping to remove about 900 tonnes of CO2 from the air every year. The plant uses filters to remove the gas, then pumps it into local greenhouses, where it helps grow vegetables.

Carbon Sink captures carbon dioxide from the air using sodium carbonate and also recycles it for use in greenhouses. The added CO_2 allows the greenhouses to grow more food with less water and less fertilizer.

The Need for Power to Build a Zero-Carbon World

The 200-year rough average atmospheric lifetime of CO_2 requires that we not only stop adding CO_2 to the air, but start removing some of it.[7] To achieve near-zero emissions by 2050 we need to:

- require the use of 100% clean renewable energy all over the world going forward; this will take a staggering amount of manufacturing power (which must be zero-carbon powered) and
- deploy a massive capacity to remove CO_2 directly from the air, which also takes an enormous amount of zero-carbon power.

We cannot keep making wind turbines and solar arrays and ocean energy installations by burning coal to make the steel. So where is the power going to come from?

1. **Conservation and Efficiency**

The world can be powered by clean renewable energy leaving the fossil fuel era behind. The first and biggest power source is the lowest tech: conservation and efficiency. The International Energy Agency (IEA) estimates that about two-thirds of the primary energy that is converted to produce electricity is lost as waste heat before it reaches the end consumer. Consumers then waste at least half of their delivered energy.

The Mars candy company and Pfizer pharmaceuticals have employed sustainability officers to reduce their global environmental footprints. Mars reduced its GHGs by 25% between 2007 and 2015, and has partnered with Sumitomo Corporation to integrate wind energy. Pfizer's new plants will recycle waste water, capture rainwater, and use LED and solar-powered lighting.

Subaru is an industry leader with its zero-landfill goal. Since 2000 it has lowered waste in its Indiana factory – the size of 60 football fields – by 60%, and in 2016 it recycled 94 million pounds of material, locating vendors for much of it. The 3 Rs (reduce, reuse, recycle), though oft repeated, will not become a game changer until local communities are organized as well as Subaru, which has sent nothing to the landfill since May 4, 2004 – believing that it's just good business.

As a nation, the EPA reported in 2016 that the United States generates more solid waste than any other nation, at 4.4 pounds of waste per person per day. Of this, 55% was residential, and 45% manufacturing. Reducing household waste to near zero across the country would be a major game changer.

The growing minimalism movement is moving in this direction by practicing simplified living in the form of de-cluttering and reduced consumption, reduced wardrobes, and smaller houses.

Energy use by municipalities for pumping is estimated to be as much as 1% of the US total. Pumping systems use about 20% of all motor-driven electrical energy and have been described as the "low hanging fruit" for saving energy in industry, municipalities, and residences. An article from *Water Online* offers "10 Tips to Save Energy on Pumping Systems."

Fossil fuel energy itself is wasted, and needs to be discouraged by much higher prices for the consumer end user, while providing viable zero-carbon alternatives.

The 2017 International Renewable Energy Agency (IRENA) report[8] predicted, by 2050, a 70% reduction in CO2 emissions with two-thirds of world energy to be provided by the synergy of increased renewable energy together with the efficient recovery of waste energy. This synergy also has important environmental and societal benefits, such as lower levels of air pollution.

2. **The Good News: Zero-Carbon Clean Everlasting Energy Technologies**

Power to replace fossil fuels will also come from a host of emerging, game-changing renewable energy technologies.

In 2012, the IPCC Special Report on Renewable Energy Sources and Climate Change Mitigation found that *the total global technical potential for renewable energy is substantially higher than global energy demand,* with solar energy having the highest potential.

IRENA has a Roadmap that assesses the potential for all countries and regions to scale up to renewable energy, and keeps abreast of regional policy options for heating, cooling and transport.[9]

At the same time, according to the International Energy Agency (IEA), renewable energy in 2017 provides only 1-3% of the world energy output while fossil fuels still provide more than 80%.

How is this discrepancy to be overcome?

We have many new sources of non-fossil-fuel, safe, clean, sustainable, and everlasting energy.

These will be discussed below and include: solar voltaic, concentrated solar thermal, the solar updraft tower, onshore wind, offshore wind, ultrahigh tethered wind, geothermal (subsurface and deep), small hydro (mini and micro), and thermal ocean, tidal ocean, and wave ocean energy sources.

The only zero-carbon energy rapidly available to bridge to 100% clean global renewable energy that we recognize is fission energy, which would accelerate the total world energy conversion and could be phased out later. For more on this, including next generation reactors, please see below

Power Density

To understand how we can overcome reliance on fossil fuels, it is first necessary to distinguish between low-power density, which is supplied by electricity for "the built environment" (buildings,

The Big Three: Solar, Wind and Nuclear Power Credit: vencavolrab

streets, and household purposes) and high-power density, which is required by heavy industry and mass transport.[10]

The rapidly advancing wind, solar, and water renewables can certainly replace the world's existing electricity (low-density power), as reported by Prof. Mark Jacobson and his zero-carbon team at Stanford University in *The Solutions Project*. (We strongly support the thoroughly researched Jacobson plan published in August 2017, which limits warming to 1.5°C.) [11]

However, today's wind turbines and solar panels do not yet have the high-density capacity for smelting metals and heavy manufacturing – so the clean renewables cannot yet make themselves. Nor can they yet power heavy transportation by road, rail, air and sea, but we must get these going now.

Over the next few decades, however, it can be assumed that the renewable sources will continue to develop a much higher output efficiency, becoming equivalent to high-power density. As we will see below, concentrated solar power and geothermal energy already are high-density. Indeed the Jacobson research team finds that 139 countries could wean themselves off fossil fuels by 2050, using only wind, water, and solar power.[12]

Fortunately, two major improving technologies are making low-density power much more available, which will drive improvements to achieve high-power density for the world's heavy lifting:

- *Energy storage* is developing fast to overcome the intermittent nature of renewable energy, and
- Community electricity is being provided at scale by *renewable energy technologies*, including low-power-dense solar and wind energy, and by high-power-dense concentrated solar energy and geothermal energy.

Energy Storage

The big gamechanger for the new clean energy economy is PV (photovoltaic) plus storage.

Batteries: The 2016 World Economic Forum included among its top 10 emerging technologies Next Generation Batteries, making large-scale power storage possible. Within the past few years, new kinds of batteries have been demonstrated that deliver high enough capacity to serve whole factories and towns. These batteries, based on sodium, aluminum, or zinc, are more affordable and more scalable.

The solution for renewable energy intermittency has been solved by large, high-energy-retaining storage batteries – lithium being the leader. Vanadium redox (or flow) batteries can be enormous, large enough to supply utility-scale electricity. They consist of two parallel tanks of chemical solutions that flow adjacent to each other past a membrane, generating a charge of moving electrons. Their capacity depends simply on the size of the tanks, and they last for over 20 years. They will help to make a zero-carbon world possible.

Tesla's new battery technologies are predicted to last more than 500 miles on a single charge, with a recharge taking about as long as it takes conventional cars to gas up.

The UK and Israel have been running tests on roadways

that are wired to charge electric cars as they ride over them. One type has metal coils embedded in the asphalt that transmit energy to a receiver to charge the vehicle's battery.

Compressed air energy storage can be used on small and large scales. Ireland's Gaelectric company is building a large scale compressed-air energy storage project to be built in caverns on the Northern Irish coast.

The hydrogen fuel cell stores energy in a way that allows renewable energy to be moved around for use. It has high power density, is much cleaner and more efficient than fossil fuels, and emits only water vapor. In the hydrogen zero-carbon economy the gas is manufactured by solar or other renewable forms of energy and then distributed to fuel cars and other transport systems.

In 2016, Japan's government set a target of 40,000 hydrogen fuel-cell vehicles on its roads and 160 fueling stations by 2020. Hawaii has started to install hydrogen fueling stations, producing its hydrogen onsite through electrolysis. In 2017, the Toyota Mirai, the Hyundai Tucson Fuel Cell and the Honda Clarity Fuel Cell became available by lease in the U.S.

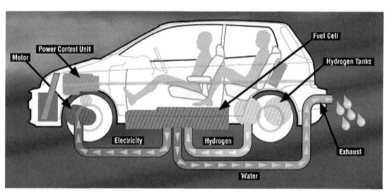

Electricity generated by hydrogen cell under the seats.
Credit: The Mellow Fellow

It should be noted that most hydrogen is produced from natural gas methane, while emitting some CO_2 in the process. A research program called Combustion of Methane without CO_2 Emissions is being carried out at the Karlsruhe Institute of Technology in Germany.

Renewable Energy Technologies

Solar Energy

Concentrated solar thermal energy has unlimited high-density potential, especially in deserts. Solar energy is collected by an array of mirrors and focused at the top of a large water tower, making steam to drive electricity-generating turbines.

In a 2011 breakthrough, the large Gemasolar plant in Seville, Spain became the first to provide uninterrupted solar thermal power for 24 hours. It consists of 2,650 heliostat mirrors that aim solar radiation at the top of a 450-foot central tower. The radiation heats molten salts, which are stored in special tanks that maintain a temperature of 500°C, allowing the plant to run steam turbines and generate electricity for up to 15 hours without any incoming solar radiation.[13]

Gemasolar concentrated solar power plant, near Seville, Spain
Credit: Marcelo Del Pozo/Reuters

In an exciting 2014 breakthrough, Australia's Commonwealth Scientific and Industrial Research Organisation generated supercritical steam in a solar thermal energy plant. This pressurized steam reached a temperature of 570°C, rivaling fossil-fuel combustion temperatures.

And in 2016, scientists at the Australian National University hit a stunning new record of 97% efficiency for converting sunlight into steam using concentrated solar power.

Solar thermal energy plants have been quick to develop. The world's largest plant, in California's Mojave Desert, is one of 60 such plants worldwide, providing electricity to 140,000 homes. In 2016, a power plant with 10,000 mirrors started operating in Nevada. In August 2017, South Australia announced construction of the world's tallest solar tower, planned to power about 5% of the region's homes.

The award-winning Magaldi (STEM) plant in Italy directs concentrated sunlight into a cylindrical steel container filled with sand, where the sand is exposed to temperatures of over 600°C. The hot sand then produces steam, which generates electricity. This technology is expected to revolutionize the solar power sector, supplying electricity and steam heat to remote zones with insufficient grids.

Ideal for concentrated solar generation, and covering a fifth of earth's surface, large deserts exist on all continents. According to the IPCC, deserts could meet all the world's power needs, and correspondingly, the industry estimates that the potential for concentrated solar is much greater than world electricity demand.

In addition, efficient transmission of electricity is now possible with advanced high voltage direct current transmission systems, integrated into smart grids, which can be well suited to local renewable energy and also to very long distance, as for concentrated solar thermal.

Thermochemical energy storage is an advance that could lead to more widespread deployment of concentrated solar power, because it can operate at significantly higher temperatures than CSP storage systems (e.g., molten salt storage).

Solar voltaic energy (photovoltaic, PV): In 1954, Bell Labs announced the invention of the first practical silicon solar cell, with a sunlight-to-electricity conversion efficiency of about 6%. In 2016, engineers at the University of New South Wales established a new PV world record efficiency of 34.5%.

The silicon dioxide used in making solar voltaic cells must be melted at 1500–2000 degrees Celsius, which is expensive and adds to CO_2 emissions. However, that is being replaced. In 2016, the World Economic Forum named the amazing perovskite solar cell among its top 10 emerging technologies.

Light-harvesting perovskite is a material that is thought to change the future of the solar cell industry. Its compound forms within certain kinds of chemical solutions. Without the need for heat, it is cheaply made into a thin light film that may be placed on virtually any surface. The PV film is then incorporated directly into building materials from windows to roofing shingles.

In September 2017, Tesla began producing solar roof tiles at its 1.2 million-square-foot factory in Buffalo, New York, making shingles that can harness the sun's energy without compromising the roof's appearance.

A 2016 research paper[14] shows a future in which solar cells will be generating energy all around us – on windows and walls, cell phones, laptops, and more.

According to IRENA, cities already account for nearly two-thirds of global energy use, and they will continue to swell.

France opens world's first solar panel road to save farmland
Credit: France 24

Solar voltaic energy may not yet be suitable for high power density, but it is increasingly planned to power not only single buildings but entire cities. By August 2017, 40 U.S. cities had committed to 100% renewable energy.

Construction started in 2016 on America's first solar-powered city. The Babcock Ranch development in Florida is designed to serve 19,500 homes, 6 million square feet of retail space, and 50,000 inhabitants. When finished in 25 years, it will produce enough energy to run the town and feed surplus power back to the grid.

Solar updraft towers: Apparently the only renewable energy technology that can generate electricity from low temperature heat is the tall solar updraft tower, where heat rises from a large warm air collector at the bottom of a tall chimney, inside of which the airflow drives wind turbines to produce electricity. This older technology requires only warm air (not direct sun) and is now being proposed for parts of Africa, Australia, and the U.S.

Geothermal Energy

The deeper we drill down in the Earth, the more geothermal energy

Clean Energy. Geothermal Power Station. Credit: Lisa-Blue

there is. This constant power-dense source could theoretically power the world. Indeed, according to a 2012 IPCC report, geothermal heat offers potential energy comparable to the global energy supply of 2008.

Countries near the Pacific Ring of Fire can access high temperatures close to the surface. In other regions, drilling down to depths of 1,000 meters or more yields geothermal temperatures of 60°C, and the temperature increases 30°C with each additional kilometer.

The granite rocks of Cornwall, England have the highest heat flow in the UK and are the best place for the development of geothermal power. The United Downs project is at the cutting edge of geothermal technology, planning to drill 4,500 meters deep to retrieve enough energy to power 5,500 homes.

The oil and gas industry has learned how to drill down to 10,000 meters, where the earth temperature is 370°C, so supply is unlimited.

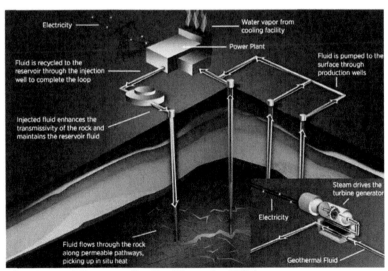

Geothermal Energy. Credit: Photonics Wiki

Geothermal facilities generate 25% of Iceland's total electricity production. Although 90 countries can access geothermal energy, only 24 of them currently produce it. This shows the vast potential for expanded geothermal energy.

Ocean Energy

Ocean Thermal Energy Conversion (OTEC) produces electricity by using the temperature difference between deep cold ocean water and warm tropical surface waters. OTEC plants pump large quantities of deep cold seawater and surface seawater to run a power cycle and produce electricity.

The deeper ocean technology can go, the more power it can generate from the gradient, which has very high potential, especially for the tropics. Most ocean energy projects are in demonstration. Large installations are expensive and the ocean power they are designed to capture can damage their structures. There are as yet only about ten successful commercially operating ocean energy installations, with about thirty projects in development.

Lockheed Martin is building the world's largest Wave energy farm, to be located off the coast of Victoria, Australia. The PowerBuoy is a piston-style wave energy harvester. Most of it is below sea level, anchored to the ocean floor. A piston is connected to a floating island – the Take Off Unit – that bobs up and down with the waves. Those movements are converted to rotational motion that spins a generator. Several PowerBuoys are connected to an Underwater Substation Pod whose output goes to shore through a subsea cable.

Tidal power is an environmentally friendly energy source, being renewable and not taking up much space. High and low tides are predictable, making it easy to construct the system with the right dimensions, knowing the flows the equipment will be exposed to.

Tidal stream generators are very similar to wind turbines. Water has 1,000 times higher density than air, which makes it possible to generate electricity at low speeds. Calculations show that power can be generated even at 1 meter per second (about 3 ft/s).

In September 2017, The Welsh Government announced £4.5 million ($5.85 million USD) in funding for marine and tidal energy development in north Wales. Meanwhile, in Orkney, tidal and wind energy are being introduced to generate hydrogen in a pilot project launched by the Scottish government, while Nova Scotia, in Canada, is also forging ahead with multiple tidal energy developments.

The SeaFlow tidal stream generator prototype with rotor raised.
Credit: Fundy

When ocean energy comes online at scale commercially, it will provide a large amount of constant, high-density power.

Nuclear Energy

The world cannot afford to close down any source of non-fossil fuel, high-energy-dense power-the densest of which by far is zero-carbon nuclear fission. The best energy plan is the Mark Jacobson Stanford plan, although recently it has been criticized as too optimistic, and its energy sources too restrictive, partly because they exclude nuclear fission.

Given the global warming emergency, the previously sensible opposition to fission is at least for now out of date. It is the wrong time to close it down. Research and development for small safe reactors should be funded, not stopped.

While coal, long in competition with nuclear fission, and historically far deadlier due to its air pollution, still remains king, small advanced compact fission reactors have safely powered large US warships since 1975, and will continue to power large new ships.

The US Navy has logged over 6,200 reactor years with no accidents. To meet the emergency response, these should be approved for land-based power (they were tested on land long ago) and built by the same contractors who built the Navy reactors under the same Energy Department testing and supervision. This could be started immediately. Back in the 1950s the U.S. Navy used liquid sodium for cooling but has since introduced a contained closed circuit water cooling system.

The best way to destroy plutonium from obsolete military nuclear warheads is to use it as nuclear fuel in a fission rector. The U.S. has imported Russia's nuclear warheads under the 20-year long *Megatons to Megawatts Program*, which ended in 2013. During this program, as much as 10% of the electricity produced in the United States was generated by fuel from Russian warheads. A small number of US warheads have been used in the same way.

Nuclear energy technology can be scaled up quickly and

is efficient and affordable for the emerging industrial economies of Asia, Africa and Latin America.

Although new reactor development has often been delayed by national safety approvals, higher stage reactors already constructed and successfully running in one country could be fast-tracked for use in other countries.

Next Generation Thorium Reactors: India has little uranium, but has the world's largest supply of thorium in its black sand. In July 2017, India announced that its fission breeder reactor was nearing completion – the world's first large scale fast breeder reactor in which thorium instead of uranium will be used as fuel.

Thorium has long held promise for safer nuclear power. A slightly radioactive element, thorium transforms into fissionable U-233 when hit by high-energy neutrons. But after use, U-233 creates fewer long-lived radioactive wastes than the conventional U-235 that is used in conventional nuclear power plants.

India's Kalpakkam reactor will generate 500 megawatts of electricity. Fast breeder reactors can extract up to 70% more energy than older fission reactors because they produce as much fuel as they consume. Breeder reactors can also use spent plutonium rods and old nuclear warheads as fuel.

This new generation reactor will also be safer because it uses a molten chemical salt as its cooling technology. It will also reduce radioactive waste several-fold. A benefit for the approval process in all countries is that thorium is harder to weaponize than uranium.

In August 2017, a Dutch nuclear research institute began work on an experimental thorium-fueled molten-salt nuclear reactor. While the West and Japan are closing down nuclear power plants, over the next two decades China plans to build the world's largest nuclear power industry in both number of power plants and advanced nuclear technology. China will develop safer next-generation reactors, including thorium molten-salt reactors, high-temperature gas-cooled reactors (which are highly efficient and inherently safe), and sodium salt-cooled reactors.

Nuclear Fusion: Inherently safe nuclear fusion, which powers the sun, is much more energy-dense than fission. Research for its development is ongoing in more than 20 locations. Recent breakthroughs, especially in artificial intelligence, suggest that commercial fusion will be possible in future decades, but it is not yet available to help solve the planetary emergency.

Tokamak Fusion Nuclear Reactor. Credit: U.S. Department of Energy

Wind Energy

A 2009 study published by the U.S. National Academy of Sciences estimated wind potential as more than 40 times the current worldwide consumption of electricity,and more than 5 times the total global use of energy in all forms.[15]

In 2015 a record 63 gigawatts of wind power were installed globally despite a dramatic drop in fossil fuel prices. Half of that was in China, while Denmark now supplies more than 40% of its electricity with wind, and Uruguay 15%.

According to the US Department of Energy, in the United States, Kansas, North Dakota and Texas could meet electricity demand for the whole country with wind power.[16]

Wind power uses 98 to 99% less water than fossil-fueled electricity, and wind farms need only one percent of the land they occupy, so are compatible with grazing, farming, and recreation. Taking just a year to build, a wind farm quickly produces energy and a return on investment. Wind fluctuations can be overcome by interconnected grids that shuttle power to where it is stored and used.

Wind turbines off Skovshoved, Denmark. Credit: Grey/Flickr

Offshore wind has become massive. In May 2017, the Netherlands opened its North Sea wind farm of 150 turbines. The largest offshore wind farm in the world is Britain's London Array of 175 turbines in the outer Thames estuary. The Burbo Bank wind farm off Liverpool was built in 2005 with 25 turbines, adding 32 turbines in 2017 to provide nearly three times its original capacity.

Wind becomes stronger at higher elevations, but taller towers are expensive to install and require much CO_2-emitting concrete.

One answer is lighter-than-air dirigibles to house the turbine. Researchers at MIT have developed a Buoyant Airborne Turbine, a giant floating generator housed inside a helium dirigible. Prototypes show that these turbines can reach altitudes of up to 600 meters, sending electricity down a tether to storage or the grid.

Google subsidiary Makani is building power-generating kites that can soar up to 300 meters – 100 meters higher than the world's largest turbine tower. The 600 kilowatt kites carry turbines aloft to generate electricity, then pass it down the tether to the grid.

Small Hydro Power

Hydroelectric power generates electricity from moving water. It requires a dependable flow of water and a reasonable height for the water to fall, where it is then fed through a pipe into a turbine. In the turbine the moving water rotates an electrical generator, which converts the motion into electrical energy.

Where water is plentiful, massive hydroelectric dams are well-known for powering major cities. "Small hydro," on the other hand, develops power on a scale suitable for local community and industry.

Small hydro is divided into mini hydro (100 to 1,000 kilowatts) and micro hydro (5 to 100 kW). Micro hydro is usually sized for smaller communities, single families or for farming. The smallest installations are pico hydro, below 5 kW.

Small hydro projects are usually fairly easy to license, and standard equipment is available, so projects may be developed quickly. Small hydro may also contribute to distributed generation in a regional electricity grid, or may be built in isolated areas where there is no grid.

In run-of-river (ROR) power, a small dam may be built on a stream to create a headpond, ensuring that there is enough water flowing through the penstock pipes to the turbines at a lower elevation. In areas such as British Columbia, a portion of a mountain stream is diverted into a buried pipe where it is channeled downstream into one or more turbines.

The sources above identify a mix of zero-carbon renewable energy that in time will develop to include plenty of high-density-power for the world's heavy-lifting industrial economy.

3. Game Changers for the Built World

Rebuilding the entire world for zero carbon in time to prevent planetary catastrophe requires much greater efficiency in the built world. The International Energy Agency reported in 2016 that the building sector uses approximately 21% of all energy produced, with much of that going into heating and cooling.[17] The first two items on the following list are not new, but are greatly under-utilized.

District Heating and Cooling (DHC): In high-density urban environments, central plants can channel hot and cool water through a network of underground pipes to many buildings, while local thermostats are independent.

Copenhagen meets 98% of its heating needs with the world's largest district system. Parallel pipes from the ocean nearby provide air conditioning.

Tokyo's DHC system has cut energy use and CO_2 emissions in half. Some university campuses use DHC, and Paris has it in the Louvre.

Heat Pumps: Heat pumps are a decades-old technology but have been widely ignored by the fossilized media. They work like refrigerators, extracting heat from the surrounding air, ground, or water, pumping it from outside to inside in winter, and reversing in summer to act as air conditioners.

They supply homes with up to five units of heat for each unit of electricity, which can be generated by renewable energy. According to the International Energy Agency, using heat pumps in 30% of buildings could reduce worldwide CO_2 emissions by 6%, an enormous contribution from a single technology.

Smart thermostats: Smart thermostats allow buildings to be heated and cooled when, where, and to the extent needed, without human intervention. They can take independent action to keep interiors comfortable while using minimal energy.

Smart glass: Windows are a major source of both heat loss and over-heating. Their efficiency has been greatly improved by invisible reflective coatings combined with insulating gas injected between the panes.

Smart glass changes color tint according to weather conditions. The tint is controlled using a smart phone as a switching device, while removing the need for blinds. Similarly, photochromic glass controls light, and thermochromic glass controls window transparency according to outside temperature.

SolarWindow has created a bendable glass veneer as thin as a business card to coat windows that will generate about 30 to 50 percent of the electricity needed for sky-scrapers, allowing an early return on the investment. A 50-story building has about six acres of glass, making the technology much more effective than rooftop panels.

Wooden Buildings: Standard steel and concrete construction, used worldwide, has a destructive impact on the climate. Concrete is made by burning limestone-emitting CO_2, and steel is made in coal furnaces.

On the other hand, as trees grow they absorb carbon, which is later stored in building materials. Dry wood is about 50% carbon, which is locked-in while the wood is in use. At the end of a building's life, its lumber can be used for other things.

Wooden buildings are a true carbon sink if designed to last hundreds of years, as the Europeans have done for centuries. A Yale study found that building with wood reduces emissions by 14% to 31% annually.[18]

Composite wood and glue lumber provides enough steel-like strength to make tall buildings. This engineered wood only chars in fires and can be pre-fabricated for inexpensive assembly onsite. Building codes, which often restrict wooden buildings to five or six stories, need to catch up with wood's improved engineering technology.

Green Roofs: The soil and vegetation on green roofs and balconies moderate indoor temperatures year-round, saving energy on both heating and air conditioning. Germany leads green roof building incentives. The Singapore government pays half the cost of green roof installation, and Chicago fast-tracks permits for green roof buildings. New roofs in San Francisco must be 15% to 30% green or use solar power, or both.

Modern residential district in Milan. Credit: RossHelen

Cool roofs reflect heat, are available in many light colors, and reduce the enormous energy used in air conditioning.

Net Zero Buildings produce as much energy as they consume all year. They exist in many countries, featuring efficient heating, cooling, and water systems, better insulation, smart windows and thermostats, and cool roofs.

The Sonnenschiff solar city in Freiburg, Germany produces four times the energy it consumes. California and Massachusetts cities require new construction to be net zero by 2020 and 2040, curbing vast amounts of wasted energy.

4. Transportation Game Changers

Transportation accounts for 23% of all global emissions.

Electric Vehicles: Worldwide, there are one billion gas-powered cars on the road, compared to one million electric vehicles. The good news is that EV sales have multiplied tenfold in a decade, and in September 2017, China announced its intention to encourage EV production by banning gas-powered cars.

If their electricity comes from fossil fuels, EVs show a 50% reduction in emissions compared to ordinary vehicles. If solar power is used, CO_2 emissions fall by 95%. Power sources for recharging must be zero-carbon; otherwise the process is defeated. "Range anxiety" is fading with 200 miles per charge now available, along with a network of cheap-to-install charging stations on the increase (such as the Tesla worldwide supercharger network), and apps to pinpoint the closest stations.

In September 2017, German automaker Daimler delivered the first few trucks to carrier UPS as part of its all-electric eCanter truck program designed for urban routes with a range of 100 kilometers (65 miles). The vehicle is powered by an electric powertrain with six high voltage lithium ion battery packs, and 500 more will be delivered by 2019 before Daimler goes into full production.

The June 15, 2017 issue of *Nature* reported a Stanford

A mock-up of an electric road (Highways England)
Credit: Smithsonian.com

University breakthrough in which it is now possible to wirelessly transmit electricity to a moving object, so EVs will be able to pick up energy from coils embedded in the highway to run on zero carbon.

Shipping: In 2015, more than 80% of global trade was moved by 90,000 commercial shipping vessels. Ships, because of their enormous size and the immense amount of fuel they consume, are among the worst polluting sources in the world. Often registered under flags that scarcely regulate them, they burn the cheapest and dirtiest oil available. The disturbing 2016 documentary, "Freightened: The Real Price of Shipping," tells the full story. Yet although shipping produces 3% of GHGs, marine vessel emissions are not included in global climate change agreements.

Solar shipping is being developed in experimental vessels. Small nuclear reactor power could be mandated for both zero-carbon freight transport and cruise ships.

In the meantime, regulations to keep shipping hulls clean (preventing drag) could save up to 40% of existing fuel costs, and regulations requiring slow steaming would lower fuel consumption by 30%.

Mass Transit: The World Bank finds that with rapid urbanization in developing countries, transportation is the fastest growing source of fossil fuel consumption and CO_2 emissions. While

Bullet train leaving Zhenjiang Station, China
Credit: Opacitatic, Wikimedia

recognizing the value of existing efficiency measures such as bus rapid transit with dedicated lanes (mostly in Latin America) and high speed rail (mostly in Asia and Europe), this chapter is mostly limited to emerging high tech solutions such as the Tesla/SpaceX experimental hyperloop. A game-changer in progress, this explicitly open-sourced design had, in September 2017, an anticipated speed of 300 mph. Similar to the Hyperloop is China's "flying train," planned to reach speeds of 1,250 mph between megacities. These trains do not use fossil fuels for propulsion, but must also be built without coal.

Aviation: Approximately 20,000 passenger and cargo airplanes are in service around the world, producing about 2.5% of global emissions.

With the exception of competition from high speed rail, there have been no breakthroughs publicized in commercial aviation to date. According to the industry, recent greatly improved efficiencies have reached their limit. Major steps must be taken to reduce non-essential air travel and to develop zero-carbon aviation.

In 2015, NASA announced its LEAPTech project, to help the aircraft industry transition to electrical propulsion within the next decade. A small private company, Bye Aerospace of Colorado, has designed and built a two-seater aircraft called the Sun Flyer (listed as "a top company to watch for 2017") that runs on electricity alone.

In September 2017, the UK-based airline EasyJet linked up with Wright Electric in the United States to build battery-powered aircraft for flights of under two hours (e.g., from London to Paris), which are expected within the decade. The U.S. has a goal to make every short flight electric within 20 years.

Lighter-than-air planes have been re-developed by Lockheed Martin for poorly accessible locations. Zeppelin is making passenger carrying airships again, while California manufacturer Aeros is building a massive airship planned for cargo. The world's largest aircraft is the lighter than air UK Airlander.

The hydrogen fuel cell is being tested for air travel, and is a natural for aircraft provided that the hydrogen is made using zero-carbon energy. This is now a practical proposition: In 2016 a breakthrough flight was made by a four-seater plane developed by the German Aerospace Center and the University of Ulm.

Working from home and buying locally: Working from the home is increasingly possible and saves up to 40% in overheads for office employers. Television and Internet entertainment and travel programs, combined with community events, can be substituted for distant travel. Already cities like Venice are protesting tourist overload.

Buying locally, which encourages cottage industries and urban gardens, holds great potential for reducing worldwide commercial shipping.

5. Methane, Nitrous Oxide, and Black Carbon Breakthroughs

Methane: Though carbon dioxide emissions are the major source of greenhouse gases, methane is far more effective at trapping heat in the atmosphere. In a warming world, since 2007 methane emitted by microscopic organisms in rice paddy fields has surged. Rice, the world's second largest staple crop, feeds about half the global population. Rice production has doubled since the 1990s to meet population growth.   The biggest source (about 35%) of

SRI system of paddy cultivation. Credit: Nithinhegde.mb

anthropogenic methane is cattle – the livestock-meat industry.

The new System of Rice Intensification (SRI) was developed in Madagascar, and is a better way of growing rice that does not use paddy fields. Demonstrated in 50 countries, the benefits of SRI include an average 60% increase in yields, up to a 90% reduction in seed, up to 50% water savings, and a 50% decrease in methane emissions.[19]

Landfills are the third largest source of methane in the United States. Biogas from landfills is about half methane and half CO_2. The methane can be captured and burned as a fuel, emitting CO_2 in the process. Newlight Technologies of California has developed a commercial catalytic process to make plastic from methane emissions. The oil-based plastic industry is a large emitter of CO_2, so making plastic out of waste methane is a two-way game changer.

Methane air capture is much more challenging than CO_2 capture because it is only present in air as parts per billion compared with parts per million with CO_2. This makes it even more vital to stop methane emissions. The big benefit of cutting methane emissions is that because methane emissions have a much shorter atmospheric lifetime than CO_2 – about 12 years – the reduction of global warming will happen faster. The most likely breakthrough may be methane-digesting bacteria, methanotrophs.

Nitrous Oxide: About 55% of N_2O emissions come from human activities, especially from burning fuels, industrial processes, and the use of nitrogen fertilizers in agriculture. Its global warming effect is 260 times that of CO_2 and emissions last in the atmosphere for 120 years. Atmospheric N_2O is increasing fast.

A new study dealing with nitrous oxide air removal was published in 2016, titled "Fighting global warming by greenhouse gas removal: destroying atmospheric nitrous oxide thanks to synergies between two breakthrough technologies."

The study proposes removing N_2O from the atmosphere using a combination of 1) the photo-catalytic breakdown of the N_2O to nitrogen and oxygen, and 2) performing this breakdown within a solar chimney power plant that generates renewable electricity. Although commercial development has not begun, this technology holds considerable promise.

Black carbon: A major game changer is the removal of black carbon (soot). This powerful greenhouse emission is not a gas. It is an aerosol or fine black particles suspended in the air, and is second only to CO_2 in absorbing heat. This means converting diesel transport to zero carbon energy and providing solar cookstoves for Africa.

Cookstoves, used by three billion people, drive deforestation, emit CO_2 and methane, and spew 25% of the world's black carbon.

Advanced biomass stoves force gases and smoke from incomplete combustion back into the stove's flame, reducing emissions by 95%.

The Global Alliance for Clean Cookstoves (GACC) aims to have 100 million clean cookstoves adopted by 2020, with universal adoption by 2030. Already 28 million households are using them.

Conclusion

The zero-carbon game changers in this short chapter offer only a small sampling of new ways to rescue our climate from fossil fuel devastation. No such sampling can remain current for long, as innovation – based mainly on market efficiency – speeds forward.

Endnotes

1   Such an approach has been described by Mark Jacobson's Solutions Project at Stanford University (http://thesolutionsproject.org).

2   The Climate Mobilization concept is discussed in: Emma Foehringer Merchant, "Should We Respond to Climate Change Like We Did to WWII?" *New Republic*, 12 May 2016.

3   Rainforest Trust (https://www.rainforesttrust.org/projects); "WLT Buy an Acre" (http://www.worldlandtrust.org/projects/buy-acre).

4   Sébastien Duyck, "Drawing Lessons for Arctic Governance from the Antarctic

Treaty System," *The Yearbook of Polar Law*, Vol. 3, 2011, 683-713

5   NBCNews, "Vegan Eating Would Slash Food's Global Warming Emissions: Study," NBC News, 12 February 2017.

6   Laurie Goering, "Carbon-sucking technology needed by 2030s, scientists warn," *Reuters*, 10 October 2017.

7   James Hansen et al. "Young people's burden: requirement of negative CO2 emissions," *Earth System. Dynam.*, 2017, 577-616 (https://www.earth-syst-dynam.net/8/577/2017/esd-8-577-2017.pdf).

8   IRENA, "Synergies between renewable energy and energy efficiency, a working paper based on REmap," International Renewable Energy Agency, 2017 (http://www.irena.org/DocumentDownloads/Publications/IRENA_REmap_Synergies_REEE_2017.pdf).

9   IRENA (http://www.irena.org/REmap/).

10   Vaclav Smil, "Power Density Primer: Understanding the Spatial Dimension of the Unfolding Transition to Renewable Electricity Generation," 8 May 2010 (http://vaclavsmil.com/wp-content/uploads/docs/smil-article-power-density-primer.pdf).

11   Mark L. Jacobson et al., "100% Clean and Renewable Wind, Water, and Sunlight All-Sector Energy Roadmaps for 139 Countries of the World," *Joule*, 23 August 2013. The Jacobson plan avoids vast numbers of premature air pollution deaths while creating millions of jobs.

12   Laura Bliss, "Charting the Planet's Path to 100% Solar Energy," *CityLab*, 23 August, 2017 (https://www.citylab.com/environment/2017/08/charting-the-planets-path-to-100-renewable-energy/537678/).

13   Torreso Energy's Gemasol Solar Power Plant (http://cspworld.org/cspworldmap/gemasolar).

14   Yi Song et al., "Visibly-Transparent Organic Solar Cells on Flexible Substrates with All-Graphene Electrodes," *Advanced Energy Materials*, 25 July 2016.

15   Xi Lu et al., "Global Potential for Wind-generated Electricity," *PNAS*, 29 April 2009 (http://www.pnas.org/content/106/27/10933.full.pdf).

16   Norman Myers and Scott Spoolman, *Environmental Issues and Solutions: A Modular Approach,* Brooks Cole, 2014, 134

17   International Energy Outlook, 2016. "Chapter 6. Buildings sector energy consumption," (https://www.eia.gov/outlooks/ieo/pdf/buildings.pdf).

18   Chadwick Oliver et al., "Carbon, Fossil Fuel, and Biodiversity Mitigation With Wood and Forests," *Journal of Sustainable Forestry*, 18 December 2013.

19   System of Rice Intensification (SRI) (http://sri.ciifad.cornell.edu/conferences/IRC2014/booth/SRI_climate_smart_rice_production_%20handout_2014.pdf).

Chapter Ten

MARKET LEADERSHIP

The Politico-Economic Context

Much has been written about the constraining effects of capitalism, globalization, and the debt-based economy on a clean energy transition, saying that we must begin by addressing these root issues.

Although these structural impediments may be slowing the potential pace of renewable energy growth, the climate emergency allows us no time to fix the economic system first.

The good news is that despite the continued criminal distortion of the markets by large government subsidies to fossil fuel corporations, renewable energy solutions seem poised to take over.

And indeed, even within the existing economic structure there has been a sea change in the energy markets since the 2015 Paris Climate Conference.

Although the Paris conference was disappointing to many scientists and climate activists for failing to legally mandate a global reduction in GHG emissions, its effect on the business sector has been remarkable.

For example, Norway's Statoil CEO, Eldar Saetre, stated in March 2017:

> Statoil is committed to developing its business in support of the ambitions of the Paris Agreement. We believe that being able to produce oil and gas with lower emissions while also growing

in profitable renewables will give competitive advantages and provide attractive business opportunities in the transition to a low-carbon economy.[1]

The Norwegian energy giant has scheduled a target to cut its carbon emissions by three million tons a year by 2030, and is planning to shift 15% to 20% of its capital spending to renewables over the next 10 years.

Following Paris, it was widely reported that 195 countries had signed an agreement to hold global temperature to less than 2°C – and if possible to 1.5°C – above pre-industrial times. This was a strong signal to the markets that policymakers intend to support aggressive decarbonizing initiatives on a global scale. To the energy sector this indicated a trend in public financing and market reforms to support the growth of renewable energy technologies around the world.

The Paris declaration has stimulated a competitive environment in which renewable energy had already been reconfiguring energy markets. While the world awaits the evolution of a more sustainable politico-economic framework, the markets, which are strongly influenced by regulatory impacts, will determine to a large degree whether humanity will deal effectively with the climate emergency or not.

The Effect of Regulatory Protections on Fossil Fuel Investors

Whatever the stance of national governments towards the climate science consensus, financial regulatory bodies such as the U.S. Securities and Exchange Commission maintain established policies requiring corporations to report adverse situations that could expose their investors to risk.

We Mean Business is a global coalition of seven international non-profits working with the world's most influential businesses to take action on climate change. In May 2017, this coalition issued a statement calling on G20 governments to formally endorse the recent climate risk disclosure recommendations put forward by the [international] Financial Stability Board's (FSB) Task-force on Climate-related Financial Disclosure (TCFD).

Late last year [2016] the TCFD group, led by Bank of England Governor Mark Carney, put forward a series of recommendations calling on listed firms to provide more detailed information on how they are responding to climate-related risks and disclose information on scenario planning, whereby they prepare for a range of different decarbonization scenarios.[2]

In March 2017, Chevron admitted that "climate factors could pose significant liability and regulatory risks to its financial returns, confirming its fears to investors that climate-related lawsuits and tighter restrictions on carbon emissions could have a significant impact on its bottom line," even rendering future oil extraction "economically infeasible."[3]

The same month, Shell executive Ben van Beurden told a group of oil experts in Houston that the industry risks losing public trust if it does not support the transition to clean energy.

This growing industry transparency is wise in light of the September 2017 denial by New York's highest court of Exxon's appeal to keep its auditor records (demonstrating awareness of climate risk) secret from a fraud investigation.[4]

But there is more widespread pressure on fossil fuels than the Paris agreement and regulatory protections.

The "Carbon Bubble"

The "carbon bubble" refers to the threat of fossil fuel assets becoming stranded as the shift to a low-carbon economy picks up speed.

Mark Campanale, founder of the independent think tank

Carbon Tracker, believes investors could be wrong to support an industry that will have to change direction within the next decade. He estimated in a July 2017 Australian ABC news story that $2 trillion worth of investments are at risk as international climate policy develops, referring to this risk as the carbon bubble.

Australian financial services regulators take climate risk to investments seriously and are showing policy leadership in this direction.

Some of Australia's biggest companies have responded by planning for a low carbon future. According to ABC News, the smart money in Australia has been veering towards investment in low-emissions industries.

Most particularly coal mining companies and their banks and investors should be protected against investing in a sector that has little future.

The carbon bubble wake-up call has been gaining momentum, and is used in rallies where black balloons representing atmospheric carbon are popped, symbolizing the collapse of the fossil fuel economy.

Has Peak Oil Demand Arrived?

It may be no accident that although we often hear that the price of oil is being suppressed by an over-supply, we do not often hear that the demand for oil is diminishing.

Increasingly, investors have become aware that funding climate change is not good business and as we have seen, even some oil companies concede that they are losing their social license.

Canada, so dependent on its oil economy, is beginning to face up to the stark reality. According to a story carried by the *Ottawa Citizen* (August 17, 2017), since mid-2014 consistently lower commodity prices have been preparing oil and gas producers for a peak demand scenario. Both large and small oil producers have lowered costs by drastically reducing their labour forces and moving toward leaner production models.

Analysts say that large multinational producers are tending to abandon megaprojects in favour of incremental growth to keep competitive in a shrinking market. This is already taking place with Imperial Oil, Husky Energy, and Suncor, who are building smaller, modest facilities.

Fossil Fuel Industry Transformation to Green Energy

Fossil fuel companies hold great potential for transformation to green energy, and given the falling demand for oil, it's in their best interest. If the solar division of French oil company Total SA were separated from its parent company, it would be one of the world's largest solar businesses. Similarly, if Norwegian oil giant Statoil moved its offshore wind business into a separate company, it would be one of the 15 largest companies on the Oslo Stock Exchange.[5]

Shifting a fossil fuel company into renewable energy can be surprisingly simple because many of the needed technical and management skills are the same. Everyone in Statoil's wind energy department, for example, was recruited internally as not much was needed to retrain its engineers.

The proceeds from "green bonds" can help oil companies to transition. The *Climate Bonds Initiative* is an international investor-focused not-for-profit "focused on mobilizing the $100 trillion bond market for climate change solutions." It helps to certify green bonds to a high standard, and has a mission "to mobilize the global bond market for climate action at scale."

Renewable Energy Market Advances

In 2016, renewable energy saw a record boost at lower cost. New solar power provided half of all new capacity, followed by wind power at one-third, and hydro-electric at 15%.

Arthouros Zervos, chair of REN21, a network of public and private sector groups covering 155 nations and 96% of the world's population, summed it up:

A global energy transition [is] well under way, with record new additions of installed renewable energy capacity, rapidly falling costs and the decoupling of economic growth and energy-related carbon dioxide emissions for the third year running.[6]

According to an August 2017 analysis by Research and Markets,[7] the Advanced Energy Storage Systems (AESS) market is projected to reach USD $19.04 billion by 2022. These storage systems include electro-chemical technology, mechanical technology, and thermal storage technology as applied to transportation and grid storage. This market is expected to expand at a compound annual growth rate (CAGR) of 8.38% from 2017 to 2022.

Another 2017 study, commissioned by the International Finance Corporation, the World Bank's ESMAP and the US Department of Energy, reports that energy storage deployment in emerging markets is expected to grow 40 percent a year over the next decade, up from today's capacity of 5 gigawatts resulting in about 80 GW of new storage capacity. This will open up new markets and offer tremendous opportunities.[8]

Regarding the global wind anemometers market, Research and Markets (R&M) has forecast growth at a CAGR of 5.57% during the period 2017-2021. One market driver is supportive regulatory policies and tax incentives for wind projects by governments worldwide, especially Germany.

R&M reports that photovoltaic glass (PV glass), used in solar cell modules, reached a global installed capacity of 76.6 GW in 2016, with a CAGR of 20.9% during 2011-2016. China accounted for 45% of total global installations.

The global photonic crystals market (which supplies the thin film coatings for windows and shingles that capture the sun's energy) is expected by R&M to reach $53,607.1 million by 2021, growing at a CAGR of 8.5% between 2017 and 2021.

Electric Vehicles to Permeate the Global Market

If major countries such as China begin buying EVs en masse, the world could see a radical shift in the demand for oil, eliminating a large part of the oil market within a few years, and vastly dimming its future prospects.

Britain, France and other European countries have already pledged to ban gasoline and diesel-based cars over the next few decades. Many others, from China to Canada, have pledged to put regulations or incentives in place to promote EV sales.

In September 2017, Volkswagen said it would invest $24 billion in zero-emissions vehicles by 2030, which would result in the unveiling of 80 new electric cars across all of its brands. That more than doubles the previous target of 30 new EV models. Also, Daimler said that its Mercedes-Benz brand would carry EV versions for all of its models by 2022.[9]

An enormous spike in electric vehicle production will mean a devastating spike in energy-dense power consumption. A major conservation role may be played by the growing electric vehicle conversion market, which keeps gas-powered cars on the road by converting them rather than building new EVs.

In March 2017, Thought.co published a list of 15 all-electric vehicle conversion companies located in the US, plus a Canadian company that converts heavy-duty diesel trucks to electric.[10] An October 2016 *Guardian* article, "Retro-electric: making petro guzzling cars eco-friendly," reported that 30,000 old cars around the world have been retro-fitted to EV by small businesses and DIY conversion kits installed by amateurs.[11] In fact, the US Department of Energy is supporting this idea.

If this upgrading function were introduced as a major division of new vehicle manufacturers, a true game changer would be born. Indeed, an Indiana plant that made the ultimate-guzzling Hummers was sold in June 2017 to SF Motors, a subsidiary of China's Sokon, in order to convert it to an electric vehicle manufacturing plant.[12]

China's Growing Share of the Renewables Market

According to the World Resources Institute, China was first in CO_2-equivalent emissions in 2013, at 25.93%, with the U.S. second at 13.87%.

Even if every other country in the world were to halt GHG emissions immediately, the climate emergency could not be headed off without China's full participation.

The Institute for Energy Economics and Financial Analysis (IEEFA) has a mission to speed the transition to a diverse, sustainable and profitable energy economy.

IEEFA director Tim Buckley wrote in a report that in 2017 China achieved global dominance in renewable energy and technology, owning five of the world's six largest solar manufacturing firms and the largest wind turbine manufacturer.[13] Buckley spoke to *The Guardian* about the report, saying the world is in the middle of a booming clean energy market, and that

> China is leaving everyone behind and has a real first-mover and scale advantage, which will be exacerbated if countries such as the US, UK and Australia continue to apply the brakes to clean energy … We are still in a relatively early stage of the transition, so the next couple of years will be defining in terms of which countries gain the major slices of the market."[14]

According to Bloomberg New Energy Finance, China was already widely recognized as the largest investor in domestic renewable energy, having invested $102 billion US in 2015 – more than twice that invested domestically by the US and about five times that of the UK.

India's Prominence in the Solar Sector

India is fifth in world CO_2-equivalent emissions at 6.43%, and is mobilizing to meet Prime Minister Narendra Modi's plans to boost solar capacity in the country to 100 GW by 2022. In 2016, India committed $3 billion in state funding to develop the country's solar panel manufacturing infrastructure.

The global investment community committed more than $100 billion to support India's solar development. Given this momentum, the path for adoption of distributed energy, off-grid systems and local storage options is becoming clear. India is also showing leadership in green finance, with its first green bank and support for local banking options through unique public-private mechanisms.[15]

Individual Renewable Energy Projects

The prestigious Energy Globe Award was founded in 1999 by the Austrian energy pioneer Wolfgang Neumann. It is a group of awards bestowed both nationally and at the world level. As many of our environmental problems already have good, feasible solutions, its purpose is to present to a broad audience successful sustainable projects from institutions, companies and organizations.

Not only does it list its annual winners; the Energy Globe also maintains a searchable database of the best projects among the 6,000 submissions it has received since 2006.[16]

Given the thousands of businesses entering the renewable energy market, identifying the leaders is a moving target. However, Wikipedia maintains a selected "list of renewable energy companies by stock exchange."

An alternative energy online trade magazine, Alten Energy Mag. com, has listed "The 50 Most Innovative New Renewable Energy Companies," an excellent source of inventions, from Proterra electric bus to Altarock, which builds and operates geothermal power plants.

Conclusion

The renewables market will have a profound impact on how

the climate change crisis turns out. The full power of market innovation can be unleashed to solve the crisis if governments remove fossil fuel subsidies and enact a carbon fee-and-dividend system that would tax fossil fuels, reward zero-carbon solutions, and thereby drive the market towards clean energy.

Endnotes

1   "Statoil presents 2030 climate roadmap," 9 March 2017 (https://www.statoil.com/en/news/2030-climate-roadmap.html).

2   Michael Holder, "Statoil, Eni and Total wake up to carbon bubble risks," *Green Biz News,* 23 May 2017.

3   Madeleine Cuff and James Murray, "Oil giants are waking up to carbon bubble risks," *GreenBiz*, 15 March 2017.

4   Erik Larson, "Exxon Must Disclose Accounting Details in New York Climate Probe," *Bloomberg*, 12 September 2017.

5   Alt Energy Stocks Climate Bonds Team, "Fossil Fuel Companies Should Be Issuing Green Bonds," June, 2016.

6   Damian Carrington, "'Spectacular' drop in renewable energy costs leads to record global boost," *The Guardian*, 6 June 2017.

7   Research and Markets, headquartered in Dublin, Ireland, has research teams in 81 countries, has more than 450 clients from Fortune 500, and has produced 1.65 million market reports (https://www.researchandmarkets.com/).

8   *Energy Storage Trends and Opportunities in Emerging Markets,* World Bank Group, International Finance Corporation, 2017.

9   Tom Kool, "OPEC Unfazed by Falling U.S. Oil Demand," *OilPrice News*, 12 September 2017.

10   Christine & Scott Gable, "Electric Vehicle (EV) Conversion Companies," *Thought.co,* 14 March 2017.

11   U.S. Department of Energy. Alternative Fuels Data Center, "Hybrid and Plug-In Electric Vehicle Conversions" (https://www.afdc.energy.gov/vehicles/electric_conversions.html). There are many "electric vehicle conversion" videos on YouTube, some specific to certain cars, such as the Honda Civic.

12   Fred Lambert, "Former Hummer factory to become an electric car factory under SF Motors*," Electrek*, 22 June 2017.

13   Tim Buckley, "China's Global Renewables Expansion," Institute for Energy Economics and Financial Analysis, January 2017.

14   Michael Slezak, "China cementing global dominance of renewable energy and technology," *The Guardian*, 6 January 2017.

15   Jennifer Delony, "Top 10 Renewable Trends to Watch in 2017," *Renewable Energy World*, 30 December 2016.

16   Energy Globe Award, Database of International Winners and Best Submissions, 2006 to present (http://www.energyglobe.info/awards).

Chapter Eleven

CIVIL RESISTANCE STRATEGIES

"A hundred years from now, looking back,
the only question that will appear important about
the historical moment in which we now live is
the question of whether or not we did anything
to arrest climate change."
The Economist,
December 12, 2011

Introduction

We begin this chapter with a quotation from award-winning criminologist, Dr. Ronald Kramer:

> The worst effects of climate change and the harms it will impose on its victims cannot be avoided unless there are drastic and quick reductions in global greenhouse gas emissions. The energy corporations and their political and ideological supporters understand all too well that achieving these necessary reductions will require a radical reordering of the economic and political systems at the heart of the global capitalist system. This is what they are desperately fighting to avoid, and they are using every powerful tool

available to them to perpetrate this catastrophic environmental state-corporate crime. A crusade, a transformative international social movement that has as its goal the drastic reduction of greenhouse gas emissions is desperately needed ... Perhaps we criminologists can play a role in provoking this movement by [generating] moral outrage at the destructive relationship between the fossil fuel industry and states that allows catastrophic climate change and its victimization to continue unabated.[1]

To inspire moral outrage, people need to *enter and feel* the stark reality of the emergency described in this book.

But for these feelings to break through, people first need to know that they can *already be empowered* to change this reality – that they can join hundreds of thousands of other people in demanding that their governments serve the interests of humanity above the interests of corporations.

This is what we pay governments to do, and *our destinies lie in compelling our representatives to finally act for our common good.*

Their failure to do so is grounds for revolution. We are not proposing the kind of violent revolution that seizes whole populations when autocratic oppression causes visible widespread suffering (although the coming climate catastrophe will dwarf historic suffering unless prevention begins now).

No, we are proposing strategies of nonviolent action, based on a clear-eyed perception of the slow but inexorable emergency that has been scandalously allowed to unfold for decades, largely as a result of media collusion with corporate and political crime.

One media crime that should cause particular outrage occurred in July 2017, when the largest U.S. bank, JPMorgan Chase, declared that it would commit to 100% renewable energy by 2020, and to $200 billion in clean energy financing by 2025.[2] Incredibly, the world was not to know of this enormously significant major bank decision because the U.S. corporate media barely mentioned it, possibly fearing other banks would follow. Even the climate-oriented *Guardian* missed this victory against fossil fuels.

How are we to stimulate the outrage to propel a "transformative international social movement that has as its goal the drastic reduction of greenhouse gas emissions?"

Emergency Mode

"Emergency mode" is the zone that human beings enter when faced by a sudden, dangerous crisis. Psychologist Dr. Margaret Klein Salamon, co-founder of the Climate Mobilization project, says that climate activists should "use the transformative power of climate truth" to lead the public out of "normal mode" into "emergency mode," which is contagious.[3]

She writes:

> This mode of human functioning, markedly different from "normal" functioning – is characterized by *an extreme focus of attention and resources on working productively to solve the emergency.*[4]

This extreme focus has been described as an intense type of "flow state." Flow is "an optimal state of consciousness when we feel our best and perform our best," like athletes who are "in the zone."

This altered state plays a key role in our survival. There's a world of difference between the emergency of our *imagination* – a brief and scary interruption of life-under-control – and the *actual peak performance* of living through a crisis. As Franklin Roosevelt famously said, "there is nothing to fear but fear itself."

In "long emergencies" such as the world wars, the business of normal life is built into an extended emergency response. Long *moral emergency mode* is based on an existential threat such as apartheid, which prompted a struggle for justice that lasted 40 years.

A parallel response to the climate crisis has been stunted by its absence in political and media debate *as a long emergency that must be solved by transforming business as usual into a new model of living.*

Rising to emergencies has saved us from extinction for eons, and is precisely why we exist as a species today. We can still mobilize powerfully to save ourselves, using the survival tools in this chapter.

Mobilization

A Gallup poll taken in March 2016 found that 41% of US adults feel global warming will pose a "serious threat" to them during their lifetimes, and 64% said that they are worried about it.[5] But a 2016 worldwide Pew poll found that people in other countries (with more responsible media) show much greater concern.[6]

Since the Paris summit the governing and corporate sectors have continued to respond in incremental, piecemeal ways. This is *not* going to reduce emissions fast enough to avoid catastrophic warming. Nationally-coordinated emergency mobilizations are required in all countries to convert their economies from fossil fuels to clean energy.

This means that citizens of all nations must demand that their governments take immediate, comprehensive WWII-style measures to make this essential transition. President Roosevelt's use of emergency presidential powers during WWII serves as an excellent model.

To catch up with Germany's war machine, Roosevelt set stunning goals for American factories: 60,000 aircraft in 1942 rising to 125,000 in 1943, and 120,000 tanks and 55,000 anti-aircraft guns in the same time period. In 1941, more than three million cars had been manufactured in the United States. In February 1942, the government banned private automobile production and only 139 more cars were made until 1945. According to Lester

Brown, auto companies represented the largest concentration of industrial power in the world. Their factories were soon turning out three quarters of the nation's aircraft engines, half the tanks, and a third of the machine guns.

Rationing was imposed on commodities, rubber, and metals needed for the war effort. The "Victory Speed" limit was posted at 35 mph and driving for pleasure was banned. Ration books were issued by volunteers to civilians, limiting a wide range of food consumption from sugar and coffee to meat, cheese, butter, and canned goods.

To supplement rationing, Americans grew their own food in "victory gardens." By 1945, some 20 million such gardens accounted for about 40% of vegetables consumed in the U.S.

But these strategies alone could not finance a war of such magnitude. War Bonds with long-term interest were issued annually and highly advertized – not only in the United States, but in Canada and Britain – inviting the whole country to invest. People flocked to buy them.[7]

The steep "Victory Tax" on high incomes reached 94% in 1944. The most progressive tax in American history, it worked to equalize disposable income.

Civilian mobilization ended unemployment, which fell in half from 7.7 million in spring 1940 to 3.4 million in fall 1941, and to an all-time low of 700,000 in fall 1944.[8]

Solving the "long emergency" of WWII had required a *reorientation* of:

- government's normal priorities
- society's normal resource allocations, and
- individuals' gratification and self-esteem activities, by spurring them to "do their part" to engage with the emergency.

This redirection of production and life-purpose away from consumerism towards an overriding national purpose bonded people in spirit, promoted egalitarianism, and greatly stimulated the economy as a whole.

The WWII model shows how powerfully nations can rally for survival in an emergency. A full climate mobilization needs to be driven by urgent petitioning from citizens of all nations, and now.

What can citizens do?

Civil Disobedience

Civil disobedience is the deliberate peaceful violation of a law. It draws its authority from a moral conflict within the individual between the laws of the country and the laws of conscience. It is a sacrifice made by individuals or groups of people who have the courage to risk prosecution and disfavor.

A case in point is Ken Ward and the "valve turners" from the Climate Disobedience Center,[9] who for publicity closed valves on oil pipelines in order to be arrested and taken to court. At trial they pursued the "climate necessity defense" that climate change is a greater harm than civil disobedience and must therefore be prevented.[10] Although Mr. Ward was barred from using the necessity defense, his second trial in May 2017 failed to find him guilty of sabotage. Progress was made, however, when the valve turners did later gain a judgment to mount a necessity defense in October 2017.[11]

Nonviolent actions may be divided into those aimed at governments, whose regulation is ultimately responsible for energy policy, and those aimed at industry, which is vulnerable to social disapproval. The Global Nonviolent Action Database lists about a thousand actions, of which we mention a few.

Farmers, in protest against governments that favor dirty energy over future generations, could band together to roll their tractors and farm equipment into towns and city halls, laden with remnants of climate-damaged crops.

Blockades are high-profile actions to slow down public services, such as the May 2016 French CGT labor action to block France's oil refineries, fuel depots, and nuclear plants.

Action can be organized from a distance, such as the 2008 Icelandic "kitchenware revolution" against the Central Bank, during which thousands of citizens assembled for the weekly banging of pots and pans in front of Iceland's Parliament. It worked: the top bankers were sent to prison.

New schools are emerging to teach organizational skills to social activists. For example, the new Sojourner Truth School, "modeled after the civil rights era's citizenship and freedom schools, teaches skills for the Trump era."[12]

Social media has been galvanizing public opinion and bringing people out to the streets, as it did with the April 29, 2017 People's Climate March, which organized a crowd of 200,000 in Washington, DC, and which ran concurrently in 300 other locations.

When civic action has risen exponentially against government policy, state forces have watched in amazement as the phenomenon of "bandwagon mobilization" unfolded.[13]

Because the cause is just, judges and juries may dilute the full strength of the law through their interpretation and sentencing, and police and military personnel may exercise lenience and/or inefficiency in disciplining offenders. Climate advocacy lawyers can provide pro bono legal services to support climate activists.

Climate Campaigns

There are three worldwide long-standing climate campaigns that people may join. These vital organizations are seldom covered by the US mainstream news or TV networks:

- 350.org, founded by Bill McKibben, uses online campaigns, grassroots organizing, and mass public actions to oppose new coal, oil and gas projects, take money out of fossil fuel companies, and build 100% clean energy solutions. It is active in 188 countries.
- The Citizens Climate Lobby (CCL), founded by Dr. James Hansen, works to empower U.S. citizens to contact members of Congress,

advocating a carbon fee and dividend plan to create a livable world. The CCL also supports the legal cases undertaken by Our Children's Trust.

- The Climate Reality Project, founded by Al Gore, recruits a critical mass of climate activists to accelerate the transition to a clean energy economy. It has over 10,000 Climate Reality leaders in 136 countries.

Other vital groups include the Climate Emergency Coalition (US), The Climate Mobilization (US), The Climate Coalition (UK), and The Leap (Canada). In Europe, the Health and Environment Alliance (HEAL) launched a campaign in 2016 focused on health to accelerate the end of fossil fuel subsidies.

In May 2016, the Break-Free movement ran a 12-day, 6-continent "global wave" campaign, involving over 30,000 participants, which disrupted 20 major fossil fuel projects and was widely covered in international media as the largest global civil disobedience to date.

Divestment from Fossil Fuels

As citizens our greatest power to influence a climate response comes from how we spend and invest our money.

Fossil fuels are on the run, which is why the media – so beholden to corporate power – did not cover bank giant JPMorgan Chase's move to renewables.

Led by the Global Divestment Mobilisation, divestment went global in May 2017, when thousands took action across 39 countries to divest their institutions from fossil fuels, declaring, "Standing on the sidelines is no longer an option."

Needless to say, this was not reported in the corporate news, but commondreams.org never fails. As Naomi Klein wrote in "The Shock Doctrine," "Common Dreams is a daily, glorious taste of what the dominant media could be if it was shaped by our common dreams of justice and peace."

Unless banks and investors stop financing coal companies now, the climate targets of Paris will not be met. Investors can consult the 2017 Urgewald database of 120 companies building new coal power plants to guide their divestments.[14]

Following Donald Trump's exit from the Paris Climate Accord, 15 NGOs representing 13 million supporters announced divestment from the "pipeline banks."[15]

The fight against climate change is clearly in the hands of investors. According to the 2017 "Carbon Majors Report," a full 71% of global industrial GHG emissions originate from just 100 fossil fuel companies, a third of which are traded on the markets.[16] Through subsidies these companies are priced into the market, but increasing the cost of their capital by phasing out their subsidies would price them out of the market – simply because the sun and wind are both free.

Phasing out fossil fuel subsidies while transferring them to renewable energy will allow oil companies to transition to clean energy in an orderly way. This is being promoted by the Lofoten Declaration:

> Over 340 non-governmental organisations from 67 countries have signed the Lofoten Declaration, which calls for an end to exploration and expansion of new oil, gas and coal reserves, a managed decline of the oil, coal, and gas industry, and a just transition to a safer climate future.[17]

Pressuring Banks to Defund Fossil Fuels

Online petitions urging banks to defund fossil fuel and pipeline operations such as the Dakota Access Pipeline and the Canadian tar sands may be signed at any time.

Funds can be transferred from banks to credit unions, or to the growing Aspiration Bank, with its new philosophy of environmentally ethical, transparent service to the middle class.

But the world's largest four banks are in China, and the

fifth is in Japan. The Chinese government is starting to embrace social activism against polluters, and requires 15,000 factories to release hourly emissions data,[18] but its biggest banks still drive new coal expansion overseas and require particular pressure from investors.

Urging Pension Fund Divestment

According to sustainable developer Assaad Razzouk, pension funds represent 25% of the world's total investments of about 100 trillion British pounds ($130 trillion USD). Pension funds are the largest single investment sector and their money belongs to working people.

Twenty funds represent one-fifth of the pension sector and dominate it. These 20 funds "are the influencers, the thought leaders, of the entire pyramid of [invested global] money. When they move, everybody else moves. Needless to say, the politicians also move."[19]

Climate risks are not priced anywhere in the world's 100 trillion pounds. Razzouk says that we need to ask our pension funds to price climate risk into their calculations to protect our futures. If the 20 fund managers move in this way, all the world's 100 trillion pounds in investments will also move, pricing out fossil fuels, and pricing in clean energy fueled lifestyles.

Razzouk concludes, "That's how you can change the world, because in fact your pension is your responsibility. Your pension is your opportunity and is the key to taking climate change action."[20]

Indeed, pension funds are starting to cut ties with the fossil fuel industry. Tom Sanzillo, who oversaw a $156 billion pension fund as deputy comptroller of New York State, said in 2017: "Oil and gas holdings have been producing substantially less value for pension funds [in recent years]. Pension funds are [re-examining their exposure to] oil and gas because of the environmental issues, but also because of weak financial performance over the past three years."[21]

Supporting Leadership by Indigenous Peoples

Across the globe, diverse native cultures practice unique traditions and provide environmental stewardship essential to Earth's collective well-being. Indigenous peoples manage an estimated 50-65% of the world's land,[22] and have resisted fossil fuel exploitation using many forms of nonviolent action.

One of the most effective things we can do about climate change is to support indigenous initiatives as they try to safeguard Mother Earth, by

- dramatically scaling up funding support and technical assistance to secure indigenous peoples' tenure to their traditional lands;
- compensating them for forest management practices that preserve carbon stocks;
- encouraging national governments to support indigenous forest carbon management as part of meeting their UN commitments to climate change mitigation and indigenous rights;
- appointing or electing them to environmental positions; and
- supporting their communities through NGOs such as Amazon Watch, and through legal defense funds such as Raven in Canada to protect their constitutional rights.

Local Initiatives

"Clictivism" on the Internet is not enough. We need to get out into the community and encourage anger and outrage at the level where people live. The PBS series "People's Century: Endangered Planet" (19th episode) shows how the anger vented towards authorities won the case for mothers in Minimata, Japan and the Love Canal.[23]

We need to teach our communities energy literacy, food literacy, climate literacy and denial literacy – one excellent tool is

Skeptical Science.[24] Many excellent films on climate change exist and can be advertized and shown locally with invitations to local media.

Presentations may be made to municipal councils, schools, local service clubs, senior residential facilities, and community cable TV.

Writing real letters to Big Oil, Big Banking, government, and the media is more effective than email: these physical letters need to be read, answered and filed.

Attractive well-designed climate-rescuing graffiti can be highly visible and difficult to remove.

A July 2017 Swedish study found that people's behavioral shifts could reduce emissions faster than waiting for national climate policies and widespread energy transformations. The six highest impact actions were: a plant-based diet, becoming car free, buying green energy, avoiding one round-trip transatlantic flight, and having fewer children.[25]

Social Media

Social media have transformed citizens from passive receivers of political information into active seekers, interpreters, and producers of news, research, and opinion.

These social media tools are now so powerful that governments and news agencies continually monitor and analyze their Facebook posts and Twitter streams to gauge public opinion.

Here are some ways to increase awareness of climate change and to persuade prominent people of the need to act quickly:

- Involve the media. Visit the FB pages of the major corporate media and check for climate news. Leave constructive comments related to the need for a climate mobilization and include links to other climate news, current YouTube coverage, tweets, books, and articles.

- Search Facebook and Twitter for individual media writers and commentators and post to them directly about the climate emergency.
- Search Facebook's main headings (top and latest stories, people, photos, videos, pages, places, groups, apps and events) for current climate action strategies that you can "like," share with friends or the public, or comment upon.
- Facebook posts-with-video are shared 12 times more frequently than text-only posts. Horizontally-shared YouTube material has accelerated the decline in top-down corporate TV viewership.
- Periscope is a live video streaming app for smart phones, owned by Twitter and launched in 2015. Demonstrations, marches, or nonviolent actions can be streamed via Twitter to the media, Twitter lists, or to fellow activists.

Some scientists are using the social media to urge climate mobilization. The billions of people active on social media make it the major forum in which the battle to implement a full climate mobilization will be played out around the world.

Conclusion

"The only thing necessary for the triumph of evil is for good men to do nothing." This well known statement, often attributed to Edmund Burke, has never been truer than in our time.

Dr. James Hansen is a true scientist and a courageous activist in struggling to overcome society's willful blindness. He has not only been arrested himself but has supported others arrested for climate-change civil disobedience.

People hope it will work out, but does our culture deserve hope? Have we done enough to earn it? Do we passively insist on hope as a way of passing the buck? How can we keep hope alive without doing something to make it so?

People need to hear that hope is not an action verb; *action is our only hope*. Most of the cultural and economy-wide actions that we need to take can only be implemented by governments who must be made to have the courage to stand up to corporations and others who are so ecologically illiterate that they put money, greed and profit before life and survival itself.

This means that the voting public needs to participate in civic engagement and create political will. We must understand that we have the power and intelligence to do that, and must refuse to sacrifice our children's futures.

Endnotes

1   Kramer, Ronald C. "Climate Change: A State-Corporate Crime Perspective." In: Toine Sappens et al., eds., *Environmental Crime and its Victims*, 2014, 34.

2   Joshua S. Hill, "JPMorgan Chase Commits To 100% Renewable Energy By 2020 & Facilitating $200 Billion In Clean Energy Financing By 2025," Clean Technica, 31 July 2017.

3   Margaret Klein Salamon, "Leading the Public into Emergency Mode: A New Strategy for the Climate Movement," 2016, 2-3.

4   Ibid., 4.

5   Oliver Milman, "'A tipping point': record number of Americans see global warming as threat," *The Guardian*, 18 March 2016.

6   Pew Research Center, "What the world thinks about climate change in 7 charts," 18 April 2016.

7   Ken Burns and Lynn Novick, "The War: War Production," WETA, Washington, DC, 2007.

8   Wikipedia, "United States home front during World War II," 27 June 2016.

9   Climate Disobedience Center (climatedisobedience.org).

10   The Civil Defense Project, *A World on Fire* (https://climatedefenseproject.org/resources/comic-explaining-the-climate-necessity-defense).

11   Sara Bernard, "Split Decision: Valve-Turner and Climate Activist Ken Ward Convicted on One Count," *Seattle Weekly*, 7 June 2017; Associated Press, "Minnesota Judge Allows 'Necessity Defense' in Pipeline Case," MPRNews, 17 October 2017.

12   Chuck Collins, "How to Go the Resistance Distance: Pop-Up Schools for Novice Activists," Moyers & Company, 19 June 2017.

13   Ronald A. Francisco, *Dynamics of Conflict*, Springer, 2010, 55 (http://bit.ly/2b4oZRk),

14   Coal-Exit Database, "Burning the Climate: 120 Top Coal Plant Developers" (https://www.coalexit.org).

15   Sierra Club, "15 Top Groups Divest from 'Pipeline Banks,'" *EcoWatch*, 6 June 2017.

16   The Carbon Majors Database, "CDP Carbon Majors Report 2017," July 2017.

17   Ivetta Gerasimchuk, "From Paris to Lofoten and Back: A Call for a Managed Decline of the Fossil-Fuel Industry," International Institute for Sustainable Development, 14 September 2017.

18   Michael Holtz, "China Embraces Public Activists – in Battling Pollution, " *Christian Science Monitor*, 17 March 2017.

19   Assaad Razzouk, "Climate Change movement has failed," TEDxUniversityofEdinburg, 20 April 2015 (https://www.youtube.com/watch?v=QkIV8PcAyr0).

20   Ibid.

21   Attracta Mooney, "Growing number of pension funds divest from fossil fuels," *Financial Times*, 27 April 2017.

22   Katie Reytar and Peter Veit, "Indigenous Peoples and Local Communities Are the World's Secret Weapon in Curbing Climate Change," World Resources Institute, 10 November 2016.

23   PBS, People's Century, Part 19, "Endangered Planet," 27 October 1996 (https://www.youtube.com/watch?v=wBd2GM3TZlI&list=PLuL26fXZ8eTNLLnugg 2BTyOZQ7HT-QZk4&index=19).

24   Skeptical Science, "Getting Skeptical About Global Warming Skepticism" (https://www.skepticalscience.com).

25   Eric Holthaus, "A Groundbreaking Study Outlines What You Can Do About Climate Change," *Grist*, 13 July 2017.

Chapter Twelve

MISSION IMPOSSIBLE?

The global climate change emergency deserves and requires a rapid global emergency response. We have seen that it would be criminally negligent to respond otherwise.

But can we absorb the reality of the crisis?

The normalcy bias – the belief that things will always, ultimately, return to normal – is a belief people enter into when facing a disaster. This belief causes them to underestimate both the possibility of a disaster and its possible effects.

We need to see that the normalcy bias is obstructing our view of the gathering climate disaster. Once we see that, we will be able to look past this bias to understand the very real crisis that is crying out for a global emergency response.

Is stopping this catastrophe before it's too late an impossible mission? We don't think so, but there are some big hurdles we need to overcome first. The first two sections below are a recap of what is urgently needed.

A Rapid Decline to Near Zero Greenhouse Gas Emissions

Timetable for the Required GHG Emissions Reductions

The mitigation plans from the Intergovernmental Panel on Climate Change (IPCC) and all other official climate agencies use CO_2-equivalent (CO_2-eq) emissions data, which report on the three main global warming gases: CO_2, methane and nitrous oxide.

All agencies agree that the decline in global emissions of CO_2 equivalents must be happening by 2020.

In June 2017, *Nature* published an expert article by six authors called "Three Years to Safeguard Our Climate," explaining that global emissions must and can decline by 2020.[1]

A 2016 *Nature* article, "Risk of multiple interacting tipping points should encourage rapid CO_2 emission reduction," had already called for "an immediate, massive effort to control CO_2 emissions, which are stopped by mid-century, leading to climate stabilization at less than 1.5 °C above pre-industrial levels."[2]

The World Bank climate change section posts emissions of total CO_2, methane and nitrous oxide from EDGAR (Emissions Database for Global Atmospheric Research). The CO_2 rate of increase has slowed but all three are still increasing and must start declining now, on the way to near zero.

The 2014 IPCC assessment said in its final Headline Statements that limiting warming to below 2°C (relative to pre-industrial levels) would require substantial emissions reductions over the next few decades, declining to "near zero" emissions of carbon dioxide and other long-lived greenhouse gases.

The only IPCC emissions scenario not above 2°C by 2100 is the best-case scenario (called RCP2.6). To achieve this best case, global CO_2-equivalents would need to go into decline by 2020. Global emissions of the three main GHGs must therefore decline on an immediate emergency basis now.

Restricting Warming to Under 1.5°C

Today we have global warming of just over 1°C (above pre-industrial times) and an average atmospheric CO_2 concentration of almost 407 ppm, which is climbing at an ever-increasing rate.

As explained earlier, the world is committed, or locked in, to a much higher degree of future warming than the current 1°C. According to the 2014 IPCC assessment, the commitment from constant atmospheric GHG concentrations corresponds to approximately 2°C of warming. To stop concentrations from getting worse means that in addition to dropping all emissions to near zero, some CO_2 must be removed from the atmosphere.

But geo-engineering, which is the deliberate large-scale intervention in earth's climate systems, is out of the question. In September 2017, UN Secretary General António Guterres warned about "the potential for states to resort to unsanctioned geo-engineering."[3] Belief in geo-engineering is dangerous because it encourages the idea that we can continue burning fossil fuels.

Any further delay in slashing global emissions is also out of the question. It is crucial to understand that we are *right out of time* to implement the game changers necessary for safeguarding the future, and that we must begin the mass transition to renewable energy now.

The Burning Age is Over

Climate science clearly calls for near zero emissions – and *zero emissions means zero emissions.*

Yet with the world's growing population and its energy demands, there is still enormous pressure to burn fossil fuels using a growing array of CO_2-polluting technologies.

In particular, there is the current hydraulic fracturing (fracking) boom for oil and natural gas. Fracking contaminates the soil and the groundwater, wasting incalculable water and threatening agriculture.

It is important to be clear about natural gas, which is promoted as a better, cleaner fossil fuel "bridge" to replace coal and other fossil fuels until zero-carbon energy can take over.

However, burning natural gas emits half the CO_2 of coal, and only a third less CO_2 than oil, which is still far too much.

Furthermore, natural gas is mostly methane, a greenhouse gas 70-80 times more powerful than CO_2. During the production and distribution of natural gas, so much methane leaks into the atmosphere that natural gas is about as dangerous as coal.

It is also important to be clear about "CCS," which is the

capture and sequestration (permanent storage) of CO_2 from power plants that burn fossil fuels or wood. CCS has been promoted for many years but is still only an idea, with the evidence indicating that it is not feasible. It has been argued that CCS is a waste of money given the high costs and low success rate – money that would be better invested in zero-carbon renewable technologies. CCS can certainly never get close to achieving near zero carbon, and we don't need it because all fossil fuel burning can be replaced by zero-carbon energy.[4]

"Negative Emissions" Technologies

Negative emissions technologies seek to remove from the atmosphere huge amounts of carbon dioxide that have already been emitted. Climate scientists are agreed that with atmospheric CO_2 so high, we must now develop ways to remove CO_2 ("negative emissions").

One such approach, "bioenergy plus carbon capture and storage" (or BECCS), would use fast-growing trees, switch-grass, and agricultural waste to capture carbon dioxide. These plants would be burned to generate power and then the CO_2 would supposedly be captured and buried. This rather cumbersome process has not been shown to be feasible – certainly not at any scale, and more importantly, it is not zero-carbon.[5]

Direct air extraction of CO_2 – pulling CO_2 from the air and turning it into a solid (calcium carbonate) – is a proven negative emissions technology (on a small scale) that reliably removes CO_2 from the atmosphere.

CO_2 can be sucked directly from the air and converted to a liquid or solid compound. This can be done using an already developed combination of purpose-built extractor fans and chemical removal of CO_2, but to do so at the scale required for survival will take a vast amount of zero-carbon energy and investment. This could be achieved with the safest compact nuclear fission (see below).

Technologies that suggest we have more time than we

have – by removing carbon from the atmosphere while continuing to burn fossil fuels – make no sense. To ensure our future survival we need a direct route to zero carbon. It is dangerous to burn any carbon at all.

Long-Term Carbon Storage

The best ways to increase long-term carbon storage are afforestation and regenerative organic agriculture[6] (which could include biochar). Afforestation is defined by the IPCC as the planting of new forests on lands that, historically, have not contained forests. For increasing carbon retention, old abandoned forestlands can be included. These newly forested areas have a carbon sink effect as they grow.

Regenerative organic agriculture increases sustained, long-term carbon storage in soils, but this is a lengthy process. Biochar is a form of charcoal produced by heating plant material with very low oxygen and a relatively low temperature.

Agricultural waste, which if burned would be a major source of GHG emissions, can instead be turned into biochar and used to enrich soils with carbon, as indigenous peoples in the Amazon have done. Conversion of crop residue to biochar and its utilization in agricultural land has a very high potential in mitigating climate change, which is why slash and burn agriculture should be converted to slash and char.

Major Sources of the Three Main GHGs

To turn mission impossible into mission possible, we need to know the major sources of the main GHG emissions. For our purposes, these sources can be simply and conveniently divided into two major categories: industrial-age energy production and industrial-age food production. Both categories emit all three major GHGs, in very different proportions. Both categories must and can be converted to virtually carbon-free processes.

The following graph from the IPCC 2014 assessment is

what we need to know to plan mitigation. It gives us the warming impact of emissions based on their sources over a 20-year period (not the usual 100 years). Studying this bar graph is very helpful in formulating humanity's survival plan.

20 year Temperature Impact of Emissions with Sources

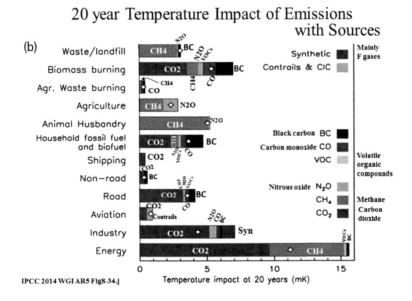

IPCC 2014 WGI AR5 Fig8-34.j   Temperature impact at 20 years (mK)

Eliminating black carbon (soot) from diesel engines, cook-stoves, and the burning of forest and savannah grasslands is a big opportunity in the fight against global climate disruption.

A 2013 review by over three dozen scientists confirmed that black carbon is second only to CO2 in contributing to global warming.[7] Cutting back soot would have an immediate effect. Other needed reductions will be discussed under our zero-combustion plan below.

Clean Energy: Growing Fast, But Far to Go

Newly-installed renewable energy capacity set new records in 2016, increasing the global total by almost 9% relative to 2015. This looks impressive, but it is not nearly fast enough. The 2016 Renewable Energy Policy Network report estimated that in 2014 the global energy share of wind, solar, geothermal and biomass combined was only 14%.[8] As we saw above, the burning of biomass is not a clean, zero-carbon energy source, which is important to acknowledge.

World energy production in 2015, as reported in the 2016 International Energy Agency's "Key World Energy Statistics," was 81.4% fossil fuels but only 1.5% clean renewable sources. The world fossil fuel share has only dropped a little from 86.7% in 1973.

World Energy Production by Fuels 1971-2015

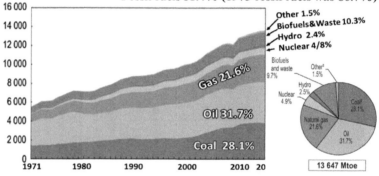

2015 Clean renewable energy production is 1.5%
Fossil fuels 81.4% (1973 fossil fuels was 86.7%)

Other Includes geothermal, solar, wind, tide/wave/ocean, heat

Source: International Energy Agency Key World Energy Statistics 2017

We cannot usher in the age of clean renewable energy while continuing to use dirty fossil fuel energy to manufacture the new energy infrastructure – which is why we need to pull off a "mission impossible" through a major global energy conversion.

The Great Conversion

Although we have far to go for a clean energy world, we have seen from the chapters *Game Changers* and *Market Leadership* that we are already entering a clean energy renaissance.

We simply need widespread recognition that the world – along with the world economy – must be rebuilt for 100% zero-combustion energy within a few years. Increasingly, this appears to be possible.

What Can Individuals Do?

The well-known 3Rs continue to be a must for climate change mitigation. In addition, we propose four top actions everyone can take specifically for the climate:

- first, learn about and talk up the emergency and crime issues and the Great Conversion proposed later in this chapter,
- second, switch all personal investments out of fossil fuels into clean renewable energy (there is good money to be made in renewables instead of bad[9]),
- third, adopt a climate- and future-friendly plant-based diet, which is also the healthiest.
- Get politically active at all government levels

But all of the above, implemented by all of us, is no longer enough.

Worldwide Mobilization

We need a massive worldwide civil mobilization to declare the planetary emergency and compel our governments to mobilize so that today's children will have a future worth living and humanity will have a future at all. Indeed this is starting to happen.

The Climate Mobilization, based in the United States, has declared a mission to protect civilization and the natural world through "a just WWII-scale mobilization that restores a safe climate." Calling for "net zero greenhouse gas emissions nationally in ten years or less and globally by 2030," this growing organization produced its "Blueprint for a Climate Emergency Movement" in 2017. [10]

Windows of Opportunity to Start the Transition Process

The current emergency has been caused by the failure of leaders and governments to take remedial action, but there are encouraging signs that things may be changing:

- The International Energy Agency reports that fossil fuel energy CO2 emissions have been flat since 2014. We at least have not increased CO2 energy emissions.
- The 1.5°C limit of the 2015 UN Paris Agreement is now the generally accepted danger limit for climate change policy-making.
- Clean energy solutions are springing up everywhere.

Given these developments, we believe that it is still possible to put global carbon emissions into rapid decline within a year, and to decarbonize the world economy in under 20 years.

As we saw earlier from an International Monetary Fund paper,[11] direct and indirect subsidies to the tune of trillions of dollars per year have kept the world dependent on fossil fuels for energy. A major opportunity for the planet's recovery is for governments to stop subsidizing fossil fuels now, which alone would put global emissions in decline within a year. This seems to be economically feasible, because in May 2016, for the first time, the G7 nations set a deadline for ending most fossil fuel subsidies, and said that government support for coal, oil and gas should end by 2025.[12]

Subsidies particularly apply to the U.S. and China, the world's highest emitting economies. China is by far the highest, with 2013 emissions at 26% of global, double those of the United States. The most optimistic current prediction is that China's emissions will not rise from that level. Climate Action Tracker ranks China's national emissions target as consistent with warming between 3°C and 4°C, which is highly insufficient.

In 2016 the Chinese government proposed a roadmap for energy subsidy reform, complete with timetables and directions.

China identified nine fossil fuel subsidies in need of reform, including subsidies for extraction and refining, for electricity and heat generation, and for end-user transportation and household consumption.[13]

In the United States, only Congress can change the subsidy laws. In both countries citizens have the opportunity to mobilize their governments to take urgent action for future survival.

There is another window for vastly improved government action within the United Nations Framework Convention on Climate Change (UNFCCC).

The May 2016 UNFCCC graph below shows the national emissions targets filed with the United Nations. It shows that global emissions must drop right away.

The graph notes the "immediate onset mitigation scenario with >66% likelihood of staying below 2°C." However the graph also shows that these current targets (called "intended nationally determined contributions," or INDCs) are far from sufficient to put the world on track to achieve 1.5°C or even 2°C. Collectively they are heading for a global suicide scenario, with emissions still increasing at 2030 and beyond.

Global GHG Emissions from National Emissions Targets
For 1.5°C (or 2C) GHG emissions decline immediately towards zero by 2050

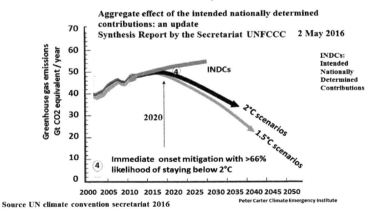

Source UN climate convention secretariat 2016

The opportunity here is that the targets undertaken by the 175 signatories to the Paris Climate Agreement are voluntary and may be greatly increased by nations at any time. This could easily be made competitive by an engaged citizenry.

Our Mission: Fast Transformation to Zero Combustion Energy and Near Zero GHG Emissions

We have an exciting challenge before us. There has never been a better time for creativity from inventors and innovators in science and technology. Indeed the game changers already invented and functioning cry out for immediate use worldwide.

The transformation of society in the graphic below can actually happen in a big way as soon as governments stop fossil fuel subsidies and impose carbon taxes instead.

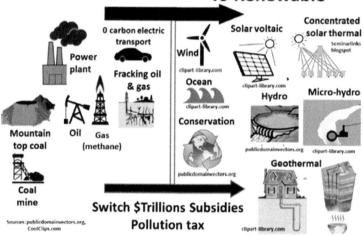

The most astute and evidence-based zero carbon plan available comes from the Stanford University team led by Mark Z. Jacobson. Titled *Meeting the World's Energy Needs Entirely with Wind, Water, and Solar Power*,[14] it is certainly the best researched, most comprehensive plan available, aiming for decarbonization of the world's energy by 2050 to avoid a long term global warming above 1.5 °C.[15]

It is based on the need for aggressive policy measures and is possibly the closest thing we have to a true zero-carbon plan for electricity, because it correctly does not include carbon capture and storage (CCS), biofuels or biomass burning.

The Solutions Project extends the Stanford wind, water and solar power plan to individual American states and countries around the world. Its roadmaps have served as a scientific basis for more than 40 cities and 100 international companies to commit to transitioning to 100% clean, renewable energy (that does not include nuclear energy). The Solutions Project provides information on all the clean renewable energy sources it covers and is our top recommendation for the global conversion to zero carbon.

Our Zero-Combustion Plan

To successfully achieve this conversion in time, we must

- use high-density zero-combustion energy to build the renewable energy infrastructure
- revolutionize food production and stop deforestation
- move agriculture off chemical fertilizers
- revolutionize building construction

We can do it all in the plan laid out below, and be much healthier and better off for it – as will all living creatures on the planet.

Use High-Density Energy

One of the biggest challenges we're facing – but not discussing – is that fossil fuel energy is high-density energy, which is ideal for powering heavy industry and heavy transportation. High density means a huge punch of energy on demand per weight of coal or volume of gasoline and natural gas.

The deadly conundrum is this: we are currently manufacturing the new renewable energy infrastructure (wind and wave turbines, solar panels and collectors) using GHG-polluting fossil fuel energy, mainly coal. Paradoxically (and perilously), our conversion to zero carbon is being powered by the dangerous combustion of carbon.

Rebuilding the world for 100% clean energy is the most monumental heavy-manufacturing project ever conceived, and will require a massive amount of high-density power. As we saw earlier, the zero-carbon renewable technologies using wind, solar and water can replace all the world's electricity but they cannot yet make themselves.

We know the conversion of all fossil fuel energy to zero-carbon energy is going to happen, with time. But right now we need a lot of high-density, zero-combustion energy with which to power the conversion to renewable energy infrastructure, and to remove CO2 directly from the air mechanically. This is our mission impossible, which we must make possible for our future survival.

This is why *nuclear energy (fission),* with its super high-density CO2-free power, is the only high-density zero-carbon bridge to build the world for 100% renewable energy. When we have power-dense renewable energy capable of smelting metal, fission can be closed down. But meanwhile it is needed: as proof, Germany's carbon emissions have increased since it closed its nuclear fission power plants.

The deployment of the safest new *compact fission reactors* can power heavy industry. Small fission reactors can certainly be used in shipping, as the US Navy has been doing for decades to power its submarines.

In 2013, climate scientist James Hansen, along with other leading climate change experts, urged governments to use nuclear power for climate change mitigation, saying that with the planet warming and carbon dioxide emissions rising fast, we cannot afford to turn away from any technology that has the potential to displace a large fraction of our carbon emissions.

As we saw in the Game Changers chapter, a great new

form of high-density, zero-carbon energy is *concentrated solar power* (CSP), which can be produced in desert areas where sun and space abound. There are many deserts around the world, making CSP a top priority for rapid deployment globally.

CSP plants are typically located near population centers. The largest CSP project in the world is in the desert near Ouarzazate, Morocco, where the power is transmitted to a nearby power station, and then on to supply a million homes in Marrakech. There are several solar power plants in the Mojave Desert that supply power to the electricity grid.

Revolutionize Food Production and Stop Deforestation
Food production is a large GHG emitter. The UN's Food and Agriculture Organization says that emissions from agriculture, fisheries and forestry nearly doubled between 1965 and 2015.[16] One estimate shows agriculture accounts for 29% of global GHG emissions,[17] not far behind energy production.

The livestock-to-meat industry and wetland rice production are leading and increasing sources of methane emissions. Global

Agriculture Sources of Greenhouse Gas Emissions

CO2 carbon dioxide
CH4 methane
N2O nitrous oxide

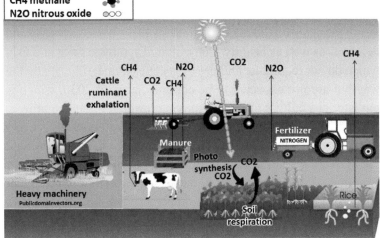

Source from EPA

warming increases the amount of methane emitted from paddy fields. Wetland (flooded) rice cultivation can be converted to dry land where possible. Research from Malaysia has shown that "modified" rice cultivation systems can increase rice yields while substantially reducing methane emissions.[18]

As cattle digest their feed they produce methane, which is then exhaled. Livestock are the greatest food production source of methane, accounting for about 30%. September 2017 saw a research paper by Dr. Julie Wolf of the U.S. Department of Agriculture (USDA), finding that methane emissions from our food production have been underestimated by over 10%.[19]

Meat production leads to tropical forest clearing, which is still going on at a fast pace and emitting CO2 through burning and soil disruption. The disturbance of soil during deforestation releases a lot of carbon as CO2, and the trees that could have re-absorbed some of that carbon are gone. Most of this is carried out for pastureland and for cattle feed.

To stop deforestation, slow the sixth mass extinction (the massive die-off of plants and animals all across the planet), and cut CO2 emissions, the world must switch from a meat-heavy "Western" diet to a plant-based one, where meat, if consumed, becomes a condiment rather than a main dish.

One of the simplest, healthiest and most effective personal conversions is to "veganize" one's diet. There are now many vegan products on the market to make a plant-based diet easy and delicious as well as healthy. Cell-cultured synthetic meat and fish, which do not require techniques of genetic engineering, are on the way but are not expected to be commercial until 2030, which is too late. The culture must start moving towards veganism now.

A carbon-intensive aspect of food growing rarely emphasized is the massive and intentional broadcast burning for agriculture that occurs in Southeast Asia, Africa and South America. These fires, which dispose of crop residues, clear areas for planting. They are incomparably larger than the worst wildfires, and emit inordinate amounts of black carbon and CO2. Getting to zero carbon means stopping this broadcast burning.

African research has shown that fire exclusion would lead to higher yields and higher ecosystem carbon stocks, while its implementation would increase labor requirements, which is seen as a large economic benefit.

Finally, the principles and practices of permaculture and regenerative agriculture can inform a conversion away from "conventional" high GHG-emitting agriculture.

Permaculture tackles how to grow food, build houses, and create communities, while minimizing environmental impact at the same time.

It is an inclusive, over-arching design system grounded in ecological principles and based on three core ethics: care for the Earth, care for others, and care for the future. Its ethics direct us to create abundance, share it fairly, and limit overconsumption to benefit the whole.

Climate change is a crisis of systems – ecosystems and social systems – and must be addressed systemically. Permaculture employs systems thinking to work with nature and to fit local conditions, terrain, and cultures.

Permaculture principles and practice may be used

Agriculture Conversion

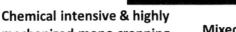

Chemical intensive & highly mechanized mono cropping

Mixed regenerative organic

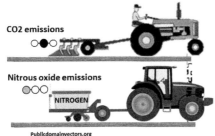

CO2 emissions

Nitrous oxide emissions

NITROGEN

Publicdomainvectors.org

Clipartlibrary.com

CO2 emissions (ploughing tillage)
Nitrous oxide emissions (chemical fertilizer)

Carbon retention
Organic fertilizer

anywhere: apartments and window boxes, suburban and country homes and gardens, allotments and smallholdings, community spaces, farms and estates, conservation areas, commercial and industrial premises, educational institutions, and waste ground. A list of permaculture strategies for dealing with the climate crisis is available online.[20]

Regenerative agriculture has been described as

> a system of farming principles and practices that increases biodiversity, enriches soils, improves watersheds, and enhances ecosystem services. It aims to capture carbon in soil and aboveground biomass, reversing current global trends of atmospheric accumulation.
>
> At the same time, it offers increased yields, resilience to climate instability, and higher health and vitality for farming and ranching communities.
>
> The system draws from decades of scientific and applied research by the global communities of organic farming, agroecology, Holistic Management, and agroforestry.[21]

Soil is the key to sequestering excess carbon. By restoring the world's degraded soils, we can store carbon as soil fertility, heal degraded land, improve water cycles and quality, and produce abundant healthy food locally.

Move Agriculture Off Chemical Fertilizers

Synthetic nitrogen fertilizer is the biggest source of nitrous oxide (NO2). The research impressively confirms that GHG emissions can be greatly reduced by more climate-friendly farming practices. In China, a 2015 farmland study showed that replacing chemical fertilizer with organic manure in temperate-climate farmland decreased the emission of GHGs, especially nitrous oxide, which reversed the agriculture ecosystem from a carbon source to a significant carbon sink.[22] Yields of wheat and corn also increased.

NO2 from fertilizers often gets overlooked as the third main greenhouse gas, although it has over 260 times the global warming power of CO2. N2O emissions last 120 years in the atmosphere, so like CO2 it is highly persistent and cumulative.
In addition, N2O is now the leading cause of stratospheric ozone depletion, so reducing its emissions to near zero is an environmental imperative.

It is essential to bring both methane and nitrous oxide down to near zero, but once again governments are paying GHG-polluting industries to keep polluting. This applies to the livestock-meat industry and to the chemically intensive mono-crop agri-business. Like fossil fuel subsidies, these subsidies must be stopped and switched to superior, non-GHG-polluting ways of producing better food.

Revolutionize Building Construction
Building with carbon-intensive steel and CO2-emitting concrete dominates major building construction worldwide. Replacing these materials with manufactured wood products will provide an enormous, often overlooked opportunity to reduce CO2 emissions. This recent re-introduction of an old method of construction, already being applied in multi-story developments, will sequester carbon for centuries.

As we make the exciting conversion to near zero greenhouse gases and the era of renewable energy, we leave our children and humanity with a truly golden future, energized forever by the sun.

The Greatest Human Venture Yet:
A Global Manhattan-Marshall-Apollo Type Plan

To implement our transition plan for zero combustion we will need to launch a massive human venture in research, development and deployment.

We could call it *Mission Earth* and think of it as a global climate-restoring Manhattan Project/Marshall Plan/Apollo type venture – inspired by the colossal undertakings that built the atom bomb in a few years, that largely rebuilt Europe after World War II in just a few years, and that put an American astronaut on the moon in 1969, less than 10 years after President Kennedy's challenge.

These monumental achievements show that our mission impossible – to restore Earth to a good and safe place where billions of people and all future generations can live and thrive – can be done.

Indeed a large international climate change venture has already been conceived and proposed. The Apollo-Gaia Project was first recommended in 2006 by David Wasdell of the UK Meridian Programme and Hans Schellnhüber, director of the Potsdam Climate Institute, following discussions with James Lovelock.

A cornerstone of the Apollo-Gaia Project is "that those in possession of the most accurate information [about climate change] share the greatest responsibility for ensuring its most effective application."[23]

Economic Issues

The Apollo-Gaia project proposal offers a project design to manage the crisis.[24] In 2015 David Wasdell updated its guidelines for implementing a global transformation to "the sustainable energy of light."[25]

In the update he identified the central economic problems of moving off fossil fuels:

> Not least we have to confront the power of addictive enslavement to fossil fuel as the energy source of our global civilisation. We tend to ignore the collateral damage. It is not just about changing our energy mix. It is not just making political decisions.
>
> There are massive profits being made from the extraction, refining, marketing and use of

fossil hydrocarbons, whether they be coal, oil, gas, fracking or tar-sands. They make the world go round. They make the money that drives the global economy. Remember that, in addition to large multi-national companies, there are some very large national economies that are totally dependent on their income from fossil energy for economic survival, for social stability, for religious coherence, and for the maintenance of political and military power.[26]

Although fossil fuels have powered the world since the Industrial Revolution, this does not necessarily mean that a transition to a new energy model would dislocate the world economy. In fact the opposite has been stated by prominent experts.

In 2014, the World Bank published the report, "Climate-smart development: Adding up the benefits of actions that help build prosperity, end poverty and combat climate change." President Jim Yong Kim commented:

Climate change poses a severe risk to global economic stability, but it doesn't have to be like this. At the World Bank Group, we believe it's possible to reduce emissions and deliver jobs and economic opportunity, while also cutting health care and energy costs.[27]

Economist Nicholas Stern, author of the 2006 UK Government's 700-page *Review on the Economics of Cimate Change,* concluded that without action the overall costs of climate change will be equivalent to losing 5% of GDP each year, possibly rising to 20% over time. In 2008, Stern estimated the annual cost of achieving climate stabilization at 2% of GDP per annum.

A 2015 study in *Nature* estimated that by 2100 unmitigated global warming will leave GDP 23% lower than without warming.

Stanford University lead-author Marshall Burke commented, "We're basically throwing away money by not addressing the issue. We see our study as providing an estimate of the benefits of reducing emissions."[28]

Indeed, global development and deployment of non-combustion technologies would be the greatest boost the world economy has ever seen. It would mean full employment worldwide and huge investment opportunities.

There are precedents for funding major emergency projects. For the Manhattan Project, the military worked with corporations and money was no object. During WWII, President Roosevelt declared an emergency and by executive orders was able to convert the entire automobile industry to making planes and tanks, using government financing. An enlightened US president could join with enlightened leaders of other nations to declare a climate emergency and order a rapid staged transition from fossil fuels to renewables.

Global Funding Sources to Rescue Earth

Funding the conversion to 100% zero-carbon energy should be the least of the problems facing our Mission Earth climate rescue.

Annual global energy investments are now about $2 trillion USD per year, with $300 billion in renewable investments accounting for less than a fifth of total energy. The U.S. defense budget for 2016 was $611 billion, more than the next eight countries combined.[29]

Two sources of funds for a Mission Earth project would be the diversion of the global $5.3 trillion (including externalities) in fossil fuel subsidies to the project, and substantial transfers from world military spending in a cooperative international effort to address the ultimate problem of climate change.

Expensive outer space explorations could be suspended as far less important (for now) than our Earth in crisis. It is time for nations to renounce aggression and non-vital spending, and to make cooperative building of a zero-combustion world their first order of priority.

There is an enormous potential role for public-private-partnerships with the development banks, who are already funding green energy at an impressive rate. These include the European Bank for Reconstruction and Development, which launched the Green Energy Transition (GET) in 2015, and the African Development Bank, whose Climate Change Fund mobilizes billions to help its member countries withstand climate change.

In addition, international consortia have been working with governments to fund and develop major projects such as the massive Noor concentrated solar project in Morocco. This offers a model for an even larger world rescue project, or network of rescue projects.

Summary

The need to wage battle on climate change has many parallels to the Allied mobilization during World War II.

Although fighting our dependence on fossil fuels is a new kind of war, it requires the same mobilization attributes – the speed, determination, resourcefulness and mass participation that was fired up to fight WWII.

We need a massive civic mobilization calling for the following initiatives to be phased in:

1. Put global carbon, methane and nitrous oxide emissions into rapid decline
2. Reduce CO2 emissions to near zero in 20 years and all GHG emissions to virtually zero by mid-century
3. Make the great conversion to electrify the entire world with zero-combustion energy:
   a. Convert all fossil fuel energy to 100% zero-combustion energy
   b. Construct the most power-dense possible super-critical steam concentrated solar thermal arrays

    with heat storage in all suitable deserts
- c. Deploy the most efficient possible energy storage in all communities worldwide
- d. Put solar photovoltaic and wind energy on track to create high-density power for heavy industry
- e. Keep super-high-density nuclear fission energy online until (d) happens
- f. Convert our diet from meat and dairy to virtually vegan, to stop deforestation and reduce methane
- g. Convert to regenerative organic agriculture to stop using chemical fertilizers
- h. Convert steel-and-concrete construction to building with certified wood from sustainable forests, which sinks carbon instead of emitting it.

4. Implement a massive Mission Earth research, development and deployment program – a global Manhattan-Marshall-Apollo Plan – for the worldwide conversions necessary to curb the climate and oceans crisis

5. Finance the research, development and rapid conversion to 100% renewable energy by the transfer of fossil fuel subsidies, reduced military spending, public private partnerships, and international cooperation

6. Research and deploy non-burning technologies for the safe, effective removal of CO_2 directly from the air.

If the will can be generated, the world has the potential to access everything it needs to carry out this plan: the scientific understanding, the funds, and the governing institutions.

Using the focused human ingenuity, cooperation, and commitment of the great projects that preceded this one, together humanity can safeguard Earth's climate and oceans for future life.

Endnotes

1.   Christiana Figueres et al., "Three Years to Safeguard Our Climate," *Nature*, 28 June 2017.

2 Yongyang Cai et al., "Risk of multiple interacting tipping points should encourage rapid CO2 emission reduction," *Nature,* 21 March 2016.

3 Camillia Born, "'UN reformer' Guterres must do more on climate change," *Climate Change News,* 20 September 2017.

4 Peter Teffer, "Europe Holds Off on Storing CO2," *EU Observer,* 25 September 2017.

5 Richard Martin, "The Dubious Promise of Bioenergy Plus Carbon Capture," *MIT Technology Review,* 8 January 2016; Naomi E. Vaughan and Clair Gough, "Expert assessment concludes negative emissions scenarios may not deliver," *Environmental Research Letters,* 31 August 2016.

6 Michael Pollan, "It's Time to Put Carbon Back Into the Soil," *EcoWatch,* 15 December 2015.

7 T.C. Bond et al., "Bounding the Role of Black Carbon in the Climate System: A Scientific Assessment," *Journal of Geophysical Research,* 6 June 2013.

8 REN21, *Renewables 2016 Global Status Report,* 2016 (http://www.ren21.net/wp-content/uploads/2016/05/GSR_2016_Full_Report_lowres.pdf).

9 Increasingly, investment companies are focusing on going fossil fuel free. Progressive Asset Management, for example, which began in 1987, supports 350.org, the US Social Investment Forum, the Interfaith Center on Corporate Responsibility, the US Climate Action Network, Ceres, the UN Principles for Responsible Investment, and other organizations promoting a more sustainable and just global economy (http://www.pamboston.com/).

10 The Climate Mobilization, *Blueprint for a Climate Emergency Movement,* 2017 (http://www.theclimatemobilization.org/blueprint).

11 David Coady et al., *How Large are Global Energy Subsidies?* IMF, 2015.

12 Karl Mathiesen, "G7 nations pledge to end fossil fuel subsidies by 2025," *The Guardian,* 27 May 2016.

13 Liu Shuang, "China to reform fossil fuel subsidies," *China Dialogue,* 16 November 2016.

14 Mark A. Delucchi and Mark Z. Jacobson, "Meeting the world's energy needs entirely with wind, water, and solar power," *Bulletin of the Atomic Scientists,* 1 July 2013.

15 Mark Z. Jacobson et al., "100% Clean and Renewable Wind, Water, and Sunlight All-Sector Energy Roadmaps for 139 Countries of the World," *Joule 1,*108–121, 6 September 2017.

16 Food and Agriculture Organization, "*Soils Help to Combat and Adapt to Climate Change by Playing a Key Role in the Carbon Cycle,*" 2015 (http://www.fao.org/3/a-i4737e.pdf).

17 CGIAR, "Agriculture and Food Production Contribute Up to 29 Percent of Global Greenhouse Gas Emissions According to Comprehensive Research Papers," 31 October 2012.

18 Pardis Fazli and Hasfalina Che Man, "Comparison of Methane Emission from Conventional and Modified Paddy Cultivation in Malaysia," *Agriculture and Agricultural Science Procedia,* 2014.

19 "Global methane emissions from agriculture larger than reported, according to new estimates," *Phys.org,* 28 September 2017 (https://phys.org/news/2017-09-global-methane-emissions-agriculture-larger.html).

20 "Permaculture Solutions for Climate Change," (http://www.permacultureclimatechange.org).

21 "Regenerative Agriculture," (http://www.regenerativeagriculturedefinition.

com).

22   Haitao Liu et al., "Mitigating greenhouse gas emissions through replacement of chemical fertilizer with organic manure in a temperate farmland," *Science Bulletin*, March 2015. Replacing chemical fertilizer with organic manure decreased GHG emissions, which reversed the agriculture ecosystem from a carbon source to a carbon sink.

23   David Wasdell, *"The Apollo-Gaia Project" Historical Background*, 18 June 2007, 7 (http://www.apollo-gaia.org/A-GProjectDevelopment.pdf).

24   Ibid., 23-26.

25   David Wasdell, *Climate Dynamics: Facing the Harsh Reality of Now: Climate Sensitivity, Target Temperature & The Carbon Budget. Guidelines for Strategic Action*, London, Apollo-Gaia Project, September 2015 (http://www.apollo-gaia.org/harsh-realities-of-now.html).

26   Ibid., 29.

27   The World Bank, "New Study Adds Up the Benefits of Climate-Smart Development in Lives, Jobs, and GDP," 23 June 2014.

28   Justin Worland, "Climate Change Could Wreck the Global Economy," *Time*, 22 October 2015.

29   Stockholm International Peace Research Institute, Military Expenditure Database, April 2017 (http://www.pgpf.org/chart-archive/0053_defense-comparison).

EVIDENCE OF THE CLIMATE EMERGENCY

In a widely publicized 2008 public statement, climate scientist James Hansen came to the startling conclusion that the world is in a state of planetary emergency.

In this Appendix, we provide the overwhelming scientific evidence of the dire state of the greenhouse gas emergency that is affecting the climate, the land and the oceans.

The comprehensive graphs below are from the August 2016 *Bulletin of the Journal of the American Meteorological Society*, which is issued every year as BAMS. All their indicators show that the adverse effects of atmospheric GHG pollution are still increasing faster than ever.

Multiple Rapidly Deteriorating Climate Change Indicators and Direct Effects to 2016

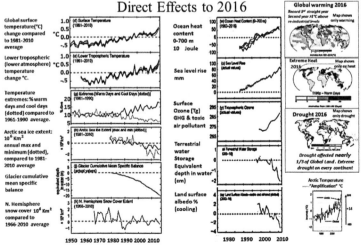

Source: BAMS 2017: State of the Climate in 2016. Bull. Amer. Meteor. Soc

Plate 1.1. Global (or representative) average time series for essential climate variables through 2016

All the impacts on the climate, land and oceans that were predicted by climate scientists years ago are now happening. The deterioration of the entire climate system has been ongoing, and has accelerated since 1980.

Why More Global Climate Change is "Locked In"

When there is more energy radiating down on the planet than there is radiating back to space, the planet is forced to heat up. "Radiative forcing" or "climate forcing" is the difference between the sunlight absorbed by the Earth and the energy radiated back to space.

Radiative forcing is the best single metric for assessing global climate change, because climate change is the direct result of the accumulated GHG emissions in the atmosphere. This is why the 1992 UNFCCC metric for climate safety is the concentration of atmospheric greenhouse gases, and not the global surface warming. NOAA's Annual Greenhouse Gas Index shows that since 1990, the radiative heat forcing of the three GHGs combined has increased by 40%.

Atmospheric CO2 Equivalent Concentration and Increased Radiative (Heat) Forcing 2016

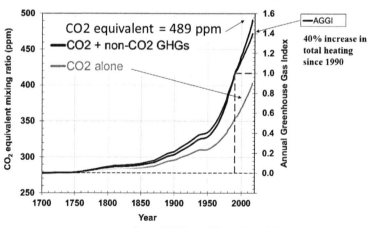

Source: THE NOAA ANNUAL GREENHOUSE GAS INDEX (AGGI)
NOAA Earth System Research Laboratory, Updated Spring 2017

Radiative forcing measures the total heat energy added to the climate system (biosphere) – about 90% of which is stored in the oceans. The time oceans take to heat up delays the eventual heating of the surface air, sea, and land. This delayed response is known as climate lag, or ocean thermal lag, and is caused by the thermal inertia of the oceans.

As we have seen, a very important thing to understand about the climate emergency is that because of ocean thermal lag, a higher degree of global warming and climate change is locked in for the future by today's atmospheric GHG concentrations. This is called "commitment," and is essential to understand. Research published by Susan Solomon et al. (*Irreversible climate change due to carbon dioxide emissions*) back in 2009 concluded that because of the ocean heat lag and the long atmospheric life time of CO2, atmospheric temperatures will not drop significantly for at least 1,000 years following the cessation of emissions.

The 2014 IPCC Assessment says that we are committed by the ocean heat lag to an additional 0.6°C by 2100, which if added to today's 1.1°C of global warming, comes to a committed temperature increase 1.7°C for 2100.

According to the 2014 Assessment, the full eventual equilibrium warming, which is the new balanced energy state after emissions stop, will take place long after 2100. At 2014 GHG concentrations, "the commitment from constant GHG concentrations would correspond to approximately 2°C of warming."[1]

The commitment calculation above does not account for emissions from the heat-boosting feedbacks (i.e., when warming releases more carbon, which leads to more warming in a vicious cycle). For example, permafrost, which holds double all the carbon in the atmosphere, is thawing under amplified surface warming in the Arctic and thereby releasing more carbon. A 2012 article published in *Nature Geoscience* estimated that permafrost feedback could result in an additional warming of up to 1.69°C by 2300. This means that the full real-world commitment as it already stands today is likely to be far more than 2°C.

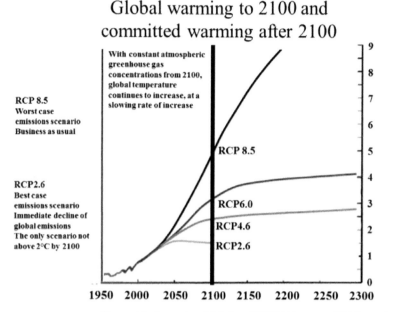

Time-evolving temperature distributions IPCC 2014 Assessment WG1 Figure 12.40

The 2014 IPCC Assessment puts it even higher. The *Working Group 3 Summary for Policy Makers* says that without additional efforts to reduce GHG emissions the warming range is 7.8°C (by 2100) when including "climate uncertainty." This uncertainty is regarded as the global feedback emissions not included in the temperature projections that IPCC models rely on.

With these known sources of additional warming, at 1.1°C today we face a risk of total biosphere collapse, requiring an immediate emergency response to put global emissions of greenhouse gases into rapid decline while reducing concentrations.

Recent Reports on the State of the Climate

Here is a summary of recent evidence showing the state of the future climate as of 2017:

- A 15-author July 2017 paper led by James Hansen says that the industrial age increase of +1°C in global average temperature "has risen well out of the Holocene range and Earth is now as warm as it was during the prior (Eemian) interglacial period, when sea level reached 6–9 m higher than today... Continued fossil fuel emissions today place a burden on young people to undertake massive technological CO2 extraction if they are to limit climate change and its consequences."[2]

- In his 2016 paper covering the Cenozoic Era, paleo-climatologist Andrew Glikson estimated that the mean post-1750 rate of global temperature rise up to 2015 is ten times faster than the previous natural record warming of the Paleocene-Eocene Thermal Maximum (PETM) 56 million years ago.
- The 2014 IPCC Assessment most importantly projects that only the best-case emissions scenario (called RCP2.6) will stay below a warming of 2°C by 2100. All other scenarios are above 2°C by 2100 and continue increasing after 2100. RCP2.6 requires global emissions to start declining now, by 2020 at the very latest. This is

Best-Case Emissions Scenario (RCP2.6) Up to 2040 (IPCC 2014)

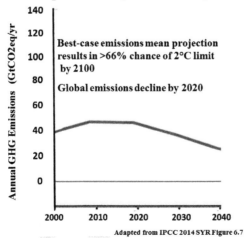

Adapted from IPCC 2014 SYR Figure 6.7

substantiated by a multi-author June 2017 *Nature* article explaining that we must "turn the tide of the world's CO2 by 2020."[3]

- The last 2016 Global Carbon Budget (for fossil fuels) and the Global Methane Budget (for all sources), published by the international expert Global Carbon Project, tells us that global CO2 and methane emissions are tracking very close to the worst-case IPCC scenario, RCP8.5.

The Escalating Arctic Emergency

Global warming, in the long days of the Arctic summer, melts the sea ice, leaving less area of white sea ice to reflect away incoming solar energy. From September 1979 to September 2016, the Arctic sea ice extent declined by 50%, as reported by the National Snow and Ice Data Center (NSIDC).

The white sea ice is replaced by dark open ocean, which has more heat-absorbing capacity than land. As a result of less energy reflection and more heat-absorbing open ocean, the Arctic warms up faster. Scientists call this Arctic amplification from the ice the albedo feedback. (The Latin word "albedo" means white, and just as white clothing is worn to reflect sunlight in hot climates, so snow and ice reflect heat from the planet in the albedo effect.) There is also the "snow albedo" feedback due to the melting away of Far North subarctic snow cover. The reduced ice and snow albedos are about equal in causing loss of global cooling effect.

In recent years, thick, old multi-year summer sea ice has dropped dramatically from the Arctic and the extent of summer sea ice cover is declining fast. The exposed ocean, being darker in color, absorbs more light and heat, causing the rate of earth's surface warming to increase. This warming causes more ice-melt, then more ocean heat absorption in what is called an amplifying or positive feedback loop.

Multiple Arctic Feedbacks

The IPCC in 2001 referred to the "runaway carbon dynamic," which is seen as a chain reaction of feedbacks, starting with the decline in Arctic summer sea ice and the loss of Far North snow cover. These two heat-boosting feedback loops are self-reinforcing and inter-reinforcing. They speed up the following Arctic feedbacks:

- warming subarctic wetlands and thawing permafrost create methane, CO_2 and nitrous oxide emissions
- boreal forest loss means CO_2 is emitted from fires and increased death of trees)
- future warming of the Arctic seas is expected to cause methane release from the Arctic subsea floor. (Solid frozen methane hydrate, a frozen gas under pressure, is a potentially massive source of methane, and is vulnerable to Arctic ocean warming. This was described by geoscientist David Archer in 2009 as "a slow but irreversible tipping point in the Earth's carbon cycle."[4)]

The temperature increase due to Arctic amplification is now running at a full and sustained 1°C above the global average, as recorded in the NOAA 2016 *Arctic Report Card* and in 2017 global temperature increases described by James Hansen at Columbia Earth Institute.

The Arctic is by far the most rapidly warming region. An April 2017 research paper revealed that Arctic permafrost is much more sensitive to warming than previously thought.[5] The authors calculate that nearly 4 million square kilometers of frozen soil would be lost for every additional degree of global warming. At 2°C above pre-industrial levels, more than 40% of existing permafrost will thaw. That would lead to catastrophic runaway Arctic feedback emissions.

Multiple Arctic Inter-reinforcing Feedbacks

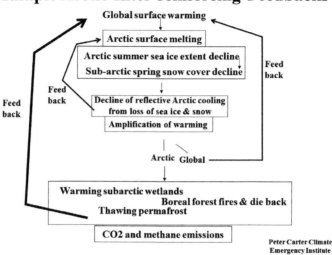

Peter Carter Climate
Emergency Institute

The Arctic is Emitting the Three Main GHGs

Emissions of methane, CO2, and nitrous oxide are being released by the rapidly warming Arctic. The melting permafrost turns to wetlands, which naturally release methane. The thawed methane creates a feedback loop that is "certain to trigger additional warming," according to the lead scientist of a 2014 research paper investigating Arctic methane emissions.[6]

The areas involved in these Arctic methane releases are enormous. In the 1980s, July Arctic sea ice was about 4 million square miles, but by 2017 covered just over 3 million square miles. Permafrost covers 5.8 million square miles, and Canadian wetlands cover about half a million square miles.

It is known that at some point the organic matter of thawing permafrost will create enough of its own heat to be irreversible. In 2013, unique field research deep in Siberian permafrost caves showed that the Siberian Arctic has an irreversible thaw-down tipping point of only 1.5°C. The paper warns that global climates only slightly warmer than today are sufficient to thaw significant regions of permafrost.[7]

The Arctic is now emitting more CO2 than expected and earlier than expected. In 2012 an international Arctic research team found that ancient coastal permafrost carbon is being rapidly converted to CO2 along 7000 km of Siberian coast as it collapses under Arctic warming.[8]

Another recent Arctic shock is that the amount of carbon dioxide emitted from northern tundra areas in the early Arctic winter between October and December has increased 70 percent since 1975. This was reported by the NOAA *2016 Arctic Report Card* and is already enough to switch the Arctic from a carbon sink to a carbon source.[9]

Several papers have reported that the Arctic is also emitting high amounts of nitrous oxide (N2O). The latest points out that there are vast stocks of nitrogen (more than 67 billion tons) in the permafrost, accumulated thousands of years ago.[10] N2O is a strong greenhouse gas, almost 300 times more powerful than CO2 for warming the climate. The researchers report that with substantial permafrost thaw N2O emissions could escape from a quarter of the Arctic surface.

As more permafrost thaws and more summer sea ice melts, the Arctic heats up with increasing speed. The possibility of a domino effect of multiple amplifying feedbacks was described by Arctic expert Carlos Duarte in 2012. The most comprehensive description of this is a review published by Fiona O'Connor and ten other authors in 2010.[11] In addition to the Arctic regional amplifying feedback, the loss of sea ice was estimated by researchers in 2014 to increase global warming by the equivalent of 25% of CO2 emissions.[12]

Evidence Arctic and Amazon Carbon Sinks Have Switched to Carbon Sources

The Arctic and the Amazon are switching from great carbon sinks to net carbon sources. Recent findings confirm our worst fears of a global warming planetary emergency.

Years ago scientists warned of the worst case scenarios that would develop if dangerous levels of global surface temperature

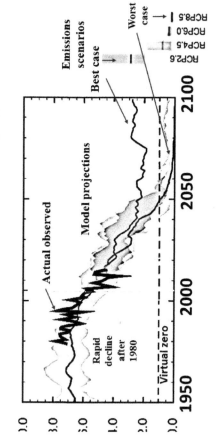

Arctic Summer Sea Ice Extent
Rapid sustained decline from 1980

a ice declines rapidly after 1980

nly the best-case scenario leaves some significant sea ice by 2100

Adapted from US Geological Survey Alaska Science Center, Climate Emergency Institute

were allowed to occur. One scenario was the switching of two major regions that normally sink carbon to instead becoming sources of emitted carbon. The two large regions that hold enormous amounts of carbon are the Arctic and the Amazon rainforest.

The Arctic: The 2016 NOAA *Arctic Report Card* had a new highlight: "Thawing permafrost releases carbon into the atmosphere, whereas greening tundra absorbs atmospheric carbon. Overall, tundra is presently releasing net carbon into the atmosphere."[13]

This would dramatically increase the rate of future global climate warming.

How the Arctic Has Switched from Carbon Sink to Carbon Source

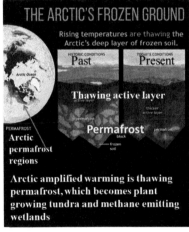

Source: 2016 NOAA Arctic Report Card

Model projections show that national emissions targets (INDCs) under the Paris Agreement will lead to an ice-free Arctic Ocean in the summer by 2050. The 2014 IPCC scenario projections show that only an immediate rapid decline in global emissions (the RCP 2.6 scenario) will leave any sea ice above the virtual zero level.

Without a climate emergency response, we can only expect the Arctic to put out increasingly more carbon emissions as the summer temperatures keep going up. This is how the runaway carbon dynamic (IPCC) starts. The only way to possibly prevent it is the immediate slashing of global GHG emissions.

The Amazon Rainforest: The carbon sink outside of the Arctic is called the terrestrial carbon sink, a very large part of which is the Amazon rainforest. A paper published back in 2003 raised the risk of Amazon die-back from continued global warming starting around mid-century, emitting CO2 and adding another 0.5°C to global warming by 2100.[14] The 2014 IPCC Assessment called the Amazon a potential tipping point.

In 2015, a *Nature* study reported a long-term decline of the Amazon carbon sink. According to this 30-year study by almost 100 researchers, the amount of carbon that the Amazon rainforest is absorbing from the atmosphere and storing each year has fallen by about a third in the last decade.[15]

In September 2017, a team of scientists at Woods Hole Research Center estimated in a research paper that degradation and destruction of the Amazon forest is now a net carbon source, emitting about as much carbon as all the vehicles on the highways in the USA.[16]

The study measured changes in above-ground forest carbon across tropical America, Africa and Asia, the most threatened forests in the world, and those with the greatest ability to act as significant carbon stores. The research adds new urgency to the critical need for aggressive global and national-scale efforts to reduce greenhouse gas emissions during this critical window of opportunity to reverse the trend in emissions by halting deforestation.

On a related topic, these net CO2 emissions from the Amazon are now adding another source of ocean acidification, as well as accelerating the increase in atmospheric CO2 and contributing to today's rapid increase in global surface warming.

The world is in a life or death race to get human emissions down fast enough before these heat-boosting feedback-dynamic emissions make it impossible for global warming and ocean acidification to stop increasing.

There must be a stop to Amazon deforestation. We cannot live without the (up to now) large carbon sinks of the Amazon and the Arctic, which the world is allowing to be destroyed. Global emissions must be put into emergency decline.

The Global GHG Emissions Emergency

Although the International Energy Agency (IEA) reported in 2017 that carbon emissions from global fossil fuel energy have been flat since 2014, they did not include all CO2 emissions, leaving out sources such as deforestation, agriculture, and waste treatment. Nor does the IEA data include total greenhouse gas emissions that include methane and nitrous oxide.

Data on total carbon dioxide, methane and nitrous oxide emissions to 2014 is provided by the World Bank Climate Change section.

Global Greenhouse Gas Emissions 1970-2014

Source: World Bank from EDGAR data
(EDGAR Emissions Database for Global Atmospheric Research)

To 2014, emissions of all three GHGs are continuing to increase. In the years since 2014, the total CO2 emissions *rate* has slowed but emissions have not declined. This has happened before. The necessary interpretation is that carbon dioxide emissions are still at an all-time catastrophic high. Survival is not simply a matter of keeping emissions from rising: they must go into rapid decline.

Global Emissions Still Increasing at 2030
Global emissions must start declining immediately

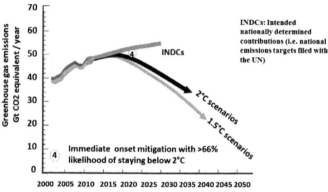

Aggregate effect of the intended nationally determined
contributions: an update
Synthesis Report by the Secretariat UNFCCC

UN Climate Secretariat May 2016

Still Accelerating CO2 Rate of Increase Has Recently Reached C02 Levels Unprecedented in Earth's History

The seasonally-adjusted mean CO2 concentration posted by NOAA in September 2017 is nearly 407 parts per million (ppm) of air (406.94). The NOAA CO2 record graph from 1960 to 2017 shows the acceleration over this period, with no sign of the increase slowing.

In 2015, the Scripps Institution of Oceanography published a short article ("Is the rate of CO2 growth slowing down or speeding up?") confirming that the rate of growth in carbon dioxide concentrations in the atmosphere has accelerated since 1960.

The growth rate of atmospheric CO2 for 1990-2000 was 1.5 ppm a year and for 2000-2010 2 ppm a year. In the 12 months from September 2016 to September 2017, it was running 2.4 ppm higher than 2016. The annual growth rates for 2015 and 2016 were unprecedented at 3.03 ppm and 2.98 ppm respectively, both boosted by the El Niño. But there was no 2017 El Niño and CO2 for 2017 is increasing substantially faster than the last recorded decade.

Atmospheric CO2 Seasonally Adjusted Mean
November 2017: 407.06 ppm
and Increasing Annual Growth Rate 2000-2016

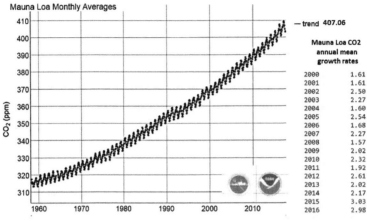

Mauna Loa CO2 annual mean growth rates	
2000	1.61
2001	1.61
2002	2.50
2003	2.27
2004	1.60
2005	2.54
2006	1.68
2007	2.27
2008	1.57
2009	2.02
2010	2.32
2011	1.92
2012	2.61
2013	2.02
2014	2.17
2015	3.03
2016	2.98

NOAA ESRL Global Monitoring Update November 2017

In its March 2016 report, NOAA described the one-year atmospheric CO2 increase of 3.05 ppm over 2015 as "explosive" and unprecedented over the ice-core record.[17] The same year, paleo-climatologist Andrew Glikson estimated that

the rate of rise in GHG over the last ~260 years, CO2 rates rising from 0.94 ppm per year in 1959 (315.97 ppm) to 1.62 ppm per year in 2000 (369.52 ppm) to 3.05 ppm per year in 2015 (400.83 ppm), constitutes a unique spike in the history of the atmosphere[18]

Message: Today's atmospheric CO2 data is more than enough to prove the dire planetary emergency. Although CO2 emissions from fossil fuel energy have been flat since 2014, with and without El Niño, atmospheric CO2 is increasing faster than ever. Without any other explanation, it looks increasingly as if the planet's greatly increased carbon sinks (due to the increase in atmospheric CO2) are starting to fail. Why is the world not being warned about this ominous turn of events? Why are just a few people still the only ones acknowledging this emergency situation?

In 2014, the world's top carbon cycle experts published "The declining uptake rate of atmospheric CO2 by land and ocean." (It was well covered in 2015 by Dr. Joe Romm in his article, "Bombshell: Land, Ocean Carbon Sinks Are Weakening, Making Climate Action More Urgent.") This takes the accelerating emergency up to a new level.

As we have seen, the world has been at a point of planetary emergency since 2008 (according to James Hansen), and for the sake of our survival, we must assume we are now passing the point of no return and we must act accordingly – which is to immediately, and with all our resources, implement a massive USA-style Manhattan-Marshall-Apollo venture, with immediate termination of fossil fuel subsidies, a hefty immediate global carbon price, and an international moratorium on deforestation. Powerful world leaders and rich high-emitting country governments must take the lead.

Methane Concentrations

Atmospheric methane is still increasing, although its rate of increase has been slowing since 2015. Methane's atmospheric global concentration, as posted by NOAA, reached 1843 parts per billion (ppb) in June 2017. Its 800,000-year limit was 800 ppm, and is now more than double its million-year high. This will increase further as Arctic methane feedback emissions increase.

Atmospheric nitrous oxide is still increasing, having reached a concentration of 330.5 ppb in September 2017.

The World Meteorological Organization *Greenhouse Gas Bulletin* for 2016 reported that from industrialization through to 2015

- atmospheric CO2 at 400 ppm had increased by 44%
- atmospheric methane at 1845 ppb had increased by 156%
- nitrous oxide at 328 ppb had increased by 21%.

Recent Atmospheric CO2 and CH4 (Methane) 2013-2017

Posted by NOAA Update November 2017

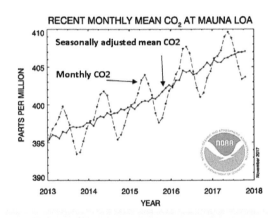

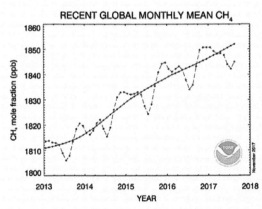

Source: National Oceanic and Atmospheric Administration, November 2017

The EPA shows that concentrations of all three main GHGs exceed their 800,000-year ice core limits. CO2's 800,000-year limit of 300 ppm is now 30% higher; methane's limit is 800 ppb and is now 130% higher.

Atmospheric Greenhouse Gas Concentrations for the Past 800,000 Years

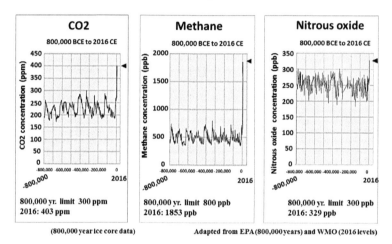

(800,000 year ice core data)　　　　Adapted from EPA (800,000 years) and WMO (2016 levels)

The Multi-Faceted Oceans Emergency

The health of the oceans is just as important to follow and understand as the integrity of the climate. The vast oceans, miles deep and wide, contain 99% of the Earth's living space and so constitute most of the climate system. Ocean health is being severely undermined by ocean heating, acidification, and deoxygenation – all a result of atmospheric greenhouse gas pollution.

The three main greenhouse gases together are causing the ocean heating, and carbon dioxide emissions cause acidification. The World Ocean is the ultimate determinant of climate and climate change. Most (95%) of the additional heat added to the biosphere by GHG emissions has gone to ocean heat.

Ocean Surface Warming Dooms Coral Reefs

Accelerating sea surface warming is causing the bleaching and death worldwide of coral reefs because the sea temperature is now too warm for the organisms to survive.

With two major extensive bleaching events in succession (2015 and 2016) damaging Australia's Great Barrier Reef, Professor Terry Hughes, director of the ARC Centre of Excellence for Coral Reef Studies, told *The Guardian* in April 2017 that the combined impact of the bleaching stretches for 1,500 kilometers (over 900 miles), leaving only the southern third unscathed. Now more than 90% of the reef has been severely damaged by global warming.[19]

Ocean Heat

Equipped with the Argo ocean sensor data, scientists have found that the oceans are warming 13% faster than previously believed.[20] Work from the Lawrence Livermore National Laboratory, Princeton University, NOAA and Penn State University shows that the heat content of the ocean has more than doubled in recent decades.[21]

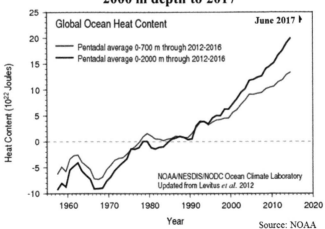

The NOAA keeps a record of the ocean heat content in both the shallower parts of the ocean (0 to 700 meters) and the deeper parts down to 2000 meters.[22] The graphs show clearly that both levels of the ocean have accelerating heat content and that from 1995 the heat accelerated faster to the deep ocean.

The 2014 IPCC Assessment projects that ocean heat content will continue to increase to 2100 under all emissions scenarios, with only the RCP2.6 best-case scenario – requiring a decline in global GHG emissions by 2020 – slowing ocean heat from 2060.

Ocean Deoxygenation

Oxygen, which is essential to ocean life, is being reduced in the oceans by global warming. First, the oxygen uptake is being lowered by the warmer water at the surface, and second, the warmer water weakens circulation to the depths and therefore less oxygen is transported downwards into the deep sea.

The first global evaluation of millions of oxygen measurements, conducted in 2017 by scientists at Helmholtz Centre for Ocean Research Kiel, showed that worldwide ocean oxygen content has decreased by more than two percent over the last 50 years, with the worst occurring in the North Pacific.[23] According to lead-author Dr. Sunke Schmidtko, these changes can have far-reaching biological consequences: for example, large fish avoid areas of low oxygen content, where they do not survive.

A 2014 IPCC study, *Ocean Systems*, led by Hans Pörtner, projects that the oceans will continue to lose oxygen, and that the rate of loss will continue under all emissions scenarios. Only in the best-case scenario (RCP2.6) does the rate of increase slow from 2060, but oxygen continues to decline to 2100. Again, the best-case scenario requires that the decline in global emissions start by 2020.

Increasing Ocean Deoxygenation and Projected Increase Under All Emissions Scenarios

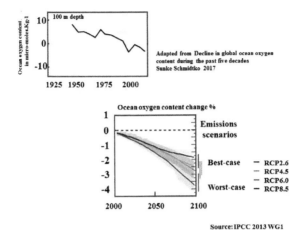

Source: IPCC 2013 WG1

Ocean Acidification

The oceans have soaked up a great deal of global CO2 emissions, causing ocean acidification. In 2015, the World Meteorological Organization published an assessment of the oceans, stating that the increase in surface ocean acidification is accelerating.

Accelerating Ocean Acidification

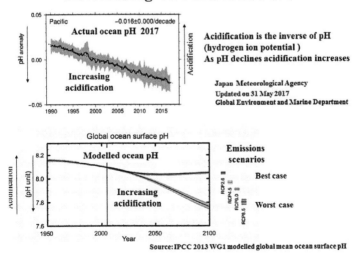

The oceans have absorbed a reported 28% of the total CO2 emissions, causing a 26% decrease in pH, which increases acidification. The WMO said "the current change appears to be the fastest in 300 million years, with the fastest known natural acidification event – occurring 55 million years ago – being probably ten times slower" than today's.[24]

The most recently available data on ocean acidification is from the Japan Meteorological Agency, updated in May 2017. This shows ocean pH in the north-western North Pacific following the accelerating trend of ocean acidification.

Marine scientist Dr. Triona McGrath warns that if we continue to burn fossil fuels at our current rate, there will be a projected 170% increase in ocean acidity by the end of this century.[25] The dissolved CO2, or carbonic acid, reduces the number of carbonate ions, which are an essential building block in seashells and coral skeletons. Over the eons, shelled ocean creatures have stored carbon as calcium carbonate in a vast carbon sink, which is now being eroded by carbonic acid.

Fisheries provide protein for at least half of the world's population. Coral reefs are home to 25% of all marine life on the planet and serve as important nurseries for fish. Ocean acidification exacerbates the deadly warming impact on corals. An 18-author study in *Nature* estimates that 25% of carbon dioxide released into the atmosphere is absorbed by the ocean. This research predicts that reefs could switch from "net accretion to net dissolution" within the century due to acidification.[26]

The 2014 IPCC Assessment projects that ocean acidification will increase under all emissions scenarios up to 2050. Only under the best-case scenario RCP2.6 does the rate of acidification slow and begin to recover after 2050. But at 2100 even the RCP2.6 acidification will be substantially greater than today.

The Sea Level Rise Emergency

The largest contribution to sea level rise over time will be the melting of planetary land ice, which in theory will continue for thousands of years.

A sobering June 2017 study in *Nature Climate Change* found that from 1993 to 2014, the rate of sea level rise increased from 2.2 millimeters per year to 3.3 millimeters – a 50% rate of increase over 20 years. The largest increase came from the Greenland ice sheet, which contributed less than 5% of all sea level rise during 1993 but more than 25% during 2014. The authors highlighted the importance and urgency of mitigating climate change and formulating coastal adaptation plans to reduce the impacts of ongoing sea-level rise.[27]

The study confirmed an April 2017 assessment that the pace of sea level rise has increased three-fold since 1990, the cause being major contributions from the Greenland and Antarctic ice sheets.[28]

Between 1992 and 2001 the Greenland ice sheet was losing 34 billion tons of ice per year, but between 2002-2011 that increased six-fold to 215 billion tons per year. Between January 2011 and December 2014 its estimated loss was an average of 269 billion tons per year of snow and ice – the equivalent of 110 million Olympic-size swimming pools worth of water each year.[29]

A study published in 2012[30] placed the lower range of the tipping point at 0.8°C, so it is possible that by 2017 (with global warming at 1.1°C) the Greenland ice sheet has passed the tipping point of near complete loss. In a stunning 2014 research paper, Antarctica made headlines in a warning that the West Antarctic ice sheet had passed a tipping point for future collapse.[31]

There is no question that significant sea level rise is already locked in for centuries, including one meter for our present century (but that could be several meters). One meter would challenge the very existence of low-lying island nations worldwide. Already, five tiny Pacific islands of the Solomon chain have disappeared due to rising seas. Six other islands had large swaths of land washed into the sea and on two of those entire villages were destroyed and people were forced to relocate.

Over 150,000 people living on the low-lying atolls of Kiribati and the Marshall Islands are threatened by rising sea levels and as flooding becomes more common relocation is the only option left. At already committed global warming and sea level rise, small low lying islands face total loss – and with no compensation. This is a serious human rights crime.

One meter of sea level rise would also dramatically increase the frequency of extreme flooding, create widespread beach and cliff erosion and property damage, making flood insurance unaffordable, and lead to salt-water intrusion in coastal aquifers.

A study in *Nature Climate Change* found that future flood losses in major coastal cities around the world may exceed $1 trillion dollars per year as a consequence of sea level rise by 2050.[32]

Long-term research shows that thermal expansion and mass loss from glaciers and the Greenland and Antarctic ice sheets will raise the sea level "ranging from 25 to 52 meters within the next 10,000 years."[33] Cities like New York, London, Rio de Janeiro, Cairo, Calcutta, Jakarta and Shanghai would all be submerged.

"The long-term perspective illustrates that policy decisions made in the next few years to decades will have profound impacts on global climate, ecosystems and human societies — not just for this century, but for the next ten millennia and beyond."[34]

Human Habitability in Danger

As global warming increases, large regions of the world will become unlivable due to maximum high summer temperatures and worsening heat waves.

A 2015 study explained that the human body may be able to adapt to extremes of low-humidity temperature through perspiration and cooling, provided that high-humidity temperature (or "mugginess") remains below 35°C.

This is the limit of survivability for a fit human being under well-ventilated outdoor conditions during high humidity heat, and is lower for most people. However, models show that high-humidity temperatures around the Arabian Gulf are likely to approach and exceed this critical 35°C under the business-as-usual scenario caused by increasing greenhouse gas concentrations.[35]

The most severe hazard from future heat waves is concentrated around densely populated agricultural regions of the Ganges and Indus river basins. Climate change presents a serious threat to human life in South Asia, which is inhabited by about a fifth of the human population.

In summary, if we are to keep the Earth's climate within the range humans are able to tolerate, we must leave the remaining fossil fuels in the ground. If we do not act now we will push the climate beyond tipping points, where the situation spirals out of our control.

The Coming Food and Water Emergency

All civilizations are totally dependent on agriculture. It is crucial that global warming be kept within the temperature range of the Holocene Period (the time of man) because this was the 10,000-year period of climate stability under which humanity was able to start and develop agriculture. Earth is now outside the temperature range that is known to be safe and conducive for agriculture.

In 2014, the IPCC made the following projections:

- There will be more frequent heat extremes over most land areas.
- These will be particularly damaging in the lower, hotter latitudes, but temperate-latitude populations and crops will also be subject to increasing frequency and severity of heat waves.[36]
- Climate change will reduce water availability in many regions, through melting glaciers and high mountain snow packs, on which huge populations depend for water.
- Climate change will reduce renewable surface

water and groundwater resources, will reduce raw water quality, and will pose risks to even treated drinking water quality.

- Without adaptation, local temperature increases in excess of about 1°C above pre-industrial could have negative effects at times on yields for the major crops (wheat, rice and maize) in both tropical and temperate regions.

- With or without adaptation, negative impacts on average annual yields for all crops in all major regions become likely from the 2030s with median yield impacts of up to 2% decline per decade projected for the rest of the century. These impacts will occur in the context of rising crop demand, which is projected to increase by about 14% per decade until 2050

These IPCC projections are summed up nicely by IRIN, a French project on food security:

One of the key effects of climate change is that extreme weather events such as floods, droughts, heatwaves, and rainfall variations become more frequent and more severe. Rising sea levels linked to climate change cause coastal erosion and loss of arable land. Rising temperatures encourage the proliferation of weeds and pests and threaten the viability of fisheries.

All this has a direct impact on agricultural production, on which the food security of most people in developing nations primarily depends. This is because agriculture in these countries is almost entirely rain-fed, and so when rains fail, or fall at the wrong time, or major storms strike, entire crops can be ruined, key infrastructure damaged or destroyed, and community assets lost. Consequently, climate change is widely seen as the greatest threat facing the estimated 500 million smallholder farmers around the world.[37]

The World Food Programme displays this message at the top of its Climate Action website:

> For millions of people across Africa, Asia and Latin America, climate change means more frequent and intense floods, droughts and storms. These can quickly spiral into full-blown food and nutrition crises. In the last decade, almost half of the World Food Programme's emergency and recovery operations have been in response to climate-related disasters, at a cost of US$23 billion.[38]

In 2016 a *Nature* study gave the Global North a wake-up call about its own food security. Droughts and extreme heat had for a time significantly reduced national cereal production by 9-10% around the world. Furthermore, the results highlight about 7% greater production damage from more recent droughts and 8-11% more damage in developed countries than in developing ones. Production levels in North America, Europe and Australasia dropped by an average of 19.9% because of droughts – roughly double the global average.[39]

The United States is not exempt. In 2009, Schlenker and Roberts assessed extreme heat in the U.S. and projected that crop yields would decrease by 30-46% at 2°C and decrease by 63-82% at 4°C.[40] The 2013 USDA report *Climate Change and Agriculture in the United States* predicted that at warming of 1-3°C, yields of major U.S. crops are projected to decline.

Another study found that climate change has significant adverse effects on production of the world's key staple crops, and could reduce global crop production by 9% by 2030, and by 23% by 2050. This constitutes a dire emergency for the whole world with respect to food security.[41]

As a result of these developments, the IPCC anticipates that violent conflicts like civil wars will become more common.

Climate Wars

The global human population is projected by the FAO to reach more than 9 billion by 2050, meaning that about 50% more food will be required. Yet as climate change and extreme weather deepen, so will lack of available food and water.

The UN reported in September 2017 that world hunger is rising again, driven by conflict and climate change, which often go hand in hand.

Following the IPCC 4th Assessment in 2007, military historian Gwynne Dyer developed six scenarios of what could happen in a global struggle for scarce resources resulting from climate-induced chaos.

According to an updated online review of Dyer's 2010 book *Climate Wars*, although the scenarios mostly take place later in the century, much of what Dyer envisaged (based on IPCC projections and the economics-based Stern Report) has started to come true already.[42]

There is no doubt that severe economic and political dislocation, followed by war and strife, will take place if the world becomes a much hungrier place.

There is also no doubt that facing the crisis now by transitioning to renewable energy as quickly as possible will be much more orderly and economical, and much less destructive and painful for humanity, than the inevitable climate wars if we fail to come to grips.

Summary

The scientific evidence for the greenhouse gas emergency, from multiple lines of evidence on climate change commitment and impacts, is overwhelming. High-emitting national governments are continuing to sacrifice our survival – and the survival of all future generations – for fossil fuel corporate profit that includes untold oil for military operations subsidized with our money.

It is time for ordinary people who understand this crisis and love their children to demand that their governments stop fossil fuel subsidies, apply a carbon tax on fossil fuels, and lead their nations in implementing the planet-healing Game Changers outlined in this book.

Endnotes

1   IPCC 2014 WG1 12.5.2.

2   James Hansen et al., "Young people's burden: requirement of negative CO2 emissions," *Earth Syst. Dynam.,* 18 July 2017.

3   Christiana Figueres et al., "Three Years to Safeguard Our Climate," *Nature*, 28 June 2017.

4   David Archer, "Ocean methane hydrates as a slow tipping point in the global carbon cycle," *PNAS,* vol. 106, no. 49, 2009.

5   S.E. Chadburn et al., "An Observation-based Constraint on Permafrost Loss as a Function of Global Warming," *Nature Climate Change*, 10 April 2017.

6   Merritt R. Turetsky et al., "A synthesis of methane emissions from 71 northern, temperate, and subtropical wetlands," *Global Change Biology*, 28 April 2014.

7   The Geological Society of London, "Siberian caves warn of permafrost meltdown," *Science Daily*, 19 June 2013.

8   J.E. Vonk et al., "Activation of old carbon by erosion of coastal and subsea permafrost in Arctic Siberia," *Nature*, 29 August 2012.

9   Róisín Commane, et al., "Carbon dioxide sources from Alaska driven by increasing early winter respiration from Arctic tundra," *PNAS*, vol. 114, no. 21, 2017.

10 C. Voigt et al., "Increased nitrous oxide emissions from Arctic peatlands after permafrost thaw," *PNAS*, 13 June 2017.

11 C.M. Duarte et al., "Tipping elements in the Arctic marine ecosystem," *Ambio*, 4 February 2012; Fiona O'Connor et al., "The Possible Role of Wetlands, Fiona M. O'Connor, "Possible role of wetlands, permafrost, and methane hydrates in the methane cycle under uuture climate change: A review," *Reviews of Geophysics*, 23 December 2010.

12 K. Pistone et al., "Observational determination of albedo decrease caused by vanishing Arctic sea ice," *PNAS*, vol. 111, no. 9, 2014.

13 NOAA Arctic Program, *Arctic Report Card Highlights, 2016* (http://www.arctic.noaa.gov/Report-Card).

14 Chris D. Jones et al., "Strong carbon cycle feedbacks in a climate model with interactive CO2 and sulphate aerosols," *Geophysical Research Letters,* vol. 30, no. 9, 2003.

15 R. J. W. Brienen et al., "Long-term decline of the Amazon carbon sink," *Nature*, 19 March 2015.

16 A. Baccini et al., "Tropical forests are a net carbon source based on aboveground measurements of gain and loss," *Science Magazine*, 28 September 2017.

17 NOAA, *Record annual increase of carbon dioxide observed at Mauna Loa for 2015*, 9 March 2016.

18	Andrew Glikson, "Cenozoic mean greenhouse gases and temperature changes with reference to the Anthropocene," *Global Change Biology*, 1 June 2016.
19	Terry Hughes, "Two-thirds of Great Barrier Reef hit by back-to-back mass coral bleaching," 1 min. video, *The Guardian*, 9 April 2017. (Maps of global ocean sea surface warming are accessible at the NASA GISS climate and temperature website.)
20	Lijing Cheng et al., "Improved estimates of ocean heat content from 1960 to 2015," *Science Magazine*, 10 March 2017.
21	Peter J. Gleckler et al., "Industrial-era global ocean heat uptake doubles in recent decades," *Nature Climate Change*, 18 January 2016.
22	NOAA, *Global Ocean Heat and Salt Content*, 23 January 2016.
23	Sunke Schmidtko et al., "Decline in global oceanic oxygen content during the past five decades." *Nature*, 2017; 542 (7641): 335
24	Toste Tanhua et al., "Monitoring Ocean Carbon and Ocean Acidification," World Meteorological Organization, *WMO Bulletin 64 (1)* 2015.
25	Triona McGrath, "Ocean acidification, global warming's evil twin, is killing the oceans," *Ocean Acidification*, 15 June 2017.
26	Rebecca Albright et al., "Reversal of ocean acidification enhances net coral reef calcification," *Nature*, 24 February 2016.
27	Xianyao Chen et al., "The increasing rate of global mean sea-level rise during 1993–2014," *Nature Climate Change*, 26 June 2017.
28	Sönke Dangendorf et al., "Reassessment of 20th century global mean sea level rise," *PNAS*, 17 April, 2017 (http://www.pnas.org/content/114/23/5946).
29	Malcolm McMillan et al., "A high-resolution record of Greenland mass balance," *Geophysical Research Letters*, 9 July 2016.
30	Alexander Robinson et al., "Multistability and critical thresholds of the Greenland ice sheet," *Nature Climate Change 2*, 11 March 2012.
31	I. Joughin et al., "Marine ice sheet collapse potentially under way for the Thwaites Glacier Basin, West Antarctica," *Science*, 16 May 2014.
32	Sean Vitousek, "What is the Cost of One Meter of Sea Level Rise?" Union of Concerned Scientists, 19 July 2017.
33	Peter U. Clark et al., "Consequences of twenty-first-century policy for multi-millennial climate and sea level change," *Nature Climate Change 6*, 8 February 2016.
34	Ibid.
35	Jeremy S. Pal and Elfatih A. B. Elfatir, "Future temperature in southwest Asia projected to exceed a threshold for human adaptability," *Nature Climate Change*, 26 October 2015.
36	IPCC 2014 5th Assessment WG1 SPM.
37	Anthony Morland, "FACT FILE: Climate Change, Food Security, and Adaptation," IRIN: The Inside Story on Emergencies, 14 June 2017.
38	World Food Programme (http://www1.wfp.org/climate-action).
39	Corey Lesk et al., "Influence of extreme weather disasters on global crop production," *Nature*, 6 January 2016.
40	Wolfram Schlenker and Michael J. Roberts, Nonlinear temperature effects indicate severe damages to U.S. crop yields under climate change," *PNAS*, vol. 108, no. 37, 2909.
41	Mekbib G. Haile et al., "Impact of Climate Change, Weather Extremes, and Price Risk on Global Food Supply, Economics of Disasters and Climate Change," June 2017.
42	Doc Snow, "Gwynn Dyer's 'Climate Wars': A Summary Review," August 2010, updated 12 May 2016.

EPILOGUE

One of the deep and enduring truths of life is that human beings are at their most sublime while immersed in the fullness and richness of the natural world. Our inborn spirits rise to meet the overpowering beauty of the eternal wilderness and the ocean.

This is what we came out of eons ago and this is our ultimate nature and truth. It takes priority in our depths over everything that modernity offers, and we are now at the place where we must fight to regain its priority in our collective decision-making.

The great spirit that inhabits all creation lies both within us and without us. Yet humanity endlessly seeks this spirit outside, as if it were somewhere beyond us.

In fact, this unlimited spirit is already present in us at all times. It is simply a matter of inviting it into our awareness as we focus on the gifts of life: the beauty of the natural world and of our families, friends and animals.

If we make a daily practice of remembering that this larger reality is both present and self-sufficient within us, we will surely be able to do what we know is right for the planet and for all who come after us.

It is from faith that this spirit is ever-available to all humanity that we have written this book.

NAME & SUBJECT INDEX